T0275951

SpringerBriefs in Molecular Science

The human Hsp60/Hsp10 complex. Snapshot based on pdb coordinates 4pj1

Peter Bross

The Hsp60 Chaperonin

 Springer

Peter Bross
Department of Clinical Medicine, Research
  Unit for Molecular Medicine
Aarhus University and Aarhus
  University Hospital
Aarhus N
Denmark

ISSN 2191-5407   ISSN 2191-5415   (electronic)
SpringerBriefs in Molecular Science
ISBN 978-3-319-26086-0   ISBN 978-3-319-26088-4   (eBook)
DOI 10.1007/978-3-319-26088-4

Library of Congress Control Number: 2015955382

Springer Cham Heidelberg New York Dordrecht London

Printed on acid-free paper

Springer International Publishing AG Switzerland is part of Springer Science+Business Media
(www.springer.com)

Preface

A still highly fascinating and perplexing scientific riddle is termed the protein folding problem: how do proteins fold into their three-dimensional structure? The molecular basis for life on earth has been greatly illuminated by breakthroughs in science in the last 70 years. We now understand to a large degree the basics of how living organisms replicate themselves, how the different types of organisms have evolved and still evolve, and also what limits and endangers the lives of organisms. We even have some clues to the human-centered enigma of aging and the limited lifespan of organisms. It is strange recapitulating that less than 100 years ago it was discovered that chromosomes are the carriers of heredity and that DNA was identified to carry the genetic information during my lifetime. While the basics of genetics are now well understood and are part of school education there are still enigmatic hurdles to capture the basis of the very pivotal last step of realization of genetic information: folding of polypeptide chains to the wide variety of three-dimensional structures that form the nanomachines and nano-devices that in biological systems perform most of the work and confer the dynamics to adapt to the environment. We know the basic 20 amino acid building blocks of polypeptides and can determine the structure of folded proteins—the nanomachines—by X-ray diffraction and a couple of other methods but we still lack a deeper understanding of how these polypeptide chains, in the short time period they get in a cellular environment, acquire the structure which confers function and renders them able to escape the degradation "police". This statement may sound overexaggerated given the large body of research and knowledge on protein folding and unfolding, on the forces determining the interactions that guide and stabilize protein structures, and the many in vitro folding and computer simulation studies of protein folding that have been accumulated in the last 50 years. However, we have now known for more than 50 years that the sequence of amino acids, at least in principle, carries all the information necessary for a protein to fold into its structure. A bunch of proteins fold correctly in the test tube. We are still puzzled that this also happens in a living cell where the conditions are very much less favorable than in the settings found to be necessary for folding in the test tube. Great enlightenment has come by the discovery of molecular chaperones—proteins themselves that 'assist' folding in the crowded cellular environment where all kinds of physicochemical disturbances for protein folding prevail. A special class of molecular chaperones, the chaperonins, is the subject of this booklet. These proteins fulfill, by virtue of their special structural organization that leads to encapsulation of proteins undergoing folding in a secluded chamber, in principle the requirements for a protein structure casting machine that shapes proteins into their structure. However, it was soon clear that chaperonins and molecular chaperones in general are 'just' facilitators of protein folding that moderate the effects of cellular disturbances for folding and help to reverse misfolding but do not confer structural information. The angle of knowledge, research activity, and speculation arising from the studies of the chaperonins —with special emphasis on the human mitochondrial chaperonin—will be discussed in this volume, trying to communicate the fascination of this subject and describing the steps on the way to a deeper understanding of protein folding this can contribute.

Contents

Contents

Abstract

This book endeavors to summarize and make available to a broad readership the current knowledge of Hsp60 chaperonins with special emphasis on the human mitochondrial Hsp60. Chaperonins represent a subgroup of the molecular chaperone family of proteins that assist and supervise other proteins in acquiring their functional conformation. Chaperonins are distinguished from other molecular chaperones by the peculiar architecture of the complexes they form that provides an inner cavity in which proteins undergoing folding are intermittenty encapsulated. Research on these proteins in the last 40 years has inspired and strongly contributed to characterize the molecular mechanisms underlying the realisation of the genetic information in living cells. This book's point of departure are the current concepts of protein folding and protein quality control in living organisms. It covers the major historic breakthroughs, evolutionary relationships, molecular mechanisms, and regulation of expression. In the final part studies on very rare inherited diseases affecting the human Hsp60 chaperonin system are discussed that have contributed novel insights into the crucial functions of the chaperonin proteins.

Keywords Hsp60 · Hsp10 · Chaperonin · Molecular chaperone · Protein quality control · Proteostasis · Protein misfolding · Protein folding · Protein misfolding diseases

Chapter 1
Introduction

Abstract Molecular chaperones are a group of proteins whose role is to assist and supervise other proteins in acquiring their cellular destination and functional conformation. They are important for cellular protein folding and protein quality control.

1.1 Molecular Chaperones and the Special Subgroup of Chaperonins

Molecular chaperones are a group of proteins whose role is to assist and supervise other proteins in acquiring their cellular destination and functional conformation. The metaphor "chaperone" is borrowed from "real life" because it intuitively lets us get some understanding of what is going on. The Merriam Webster dictionary defines the societal chaperone as "an older person who accompanies young people at a social gathering to ensure proper behavior" and the protein chaperone as "a class of proteins that facilitate the proper folding of proteins by binding to and stabilizing unfolded or partially folded proteins." So the aim—in a social human context—is proper behavior or—in the world of living cells—to help newly synthesized proteins or proteins that have partially unfolded getting into the proper/functional shape and to the right place.

Chaperonins are a subgroup of molecular chaperones that are distinguished by their oligomeric ring architecture that forms an inner cavity that can be opened and closed. They have been termed "folding machines" (Richardson et al. 1998; Saibil 2013) as they bind and encapsulate proteins that undergo folding and expose them to changing surfaces and mechanical stretching. After a certain time period of encapsulation inside the cavity, the substrate protein is released and the machine takes on the next protein. What exactly goes on in the folding chamber is still a subject of research. But it appears that it offers a series of physicochemical options for a protein undergoing folding that allows it to get ahead in its folding to the native state. By providing an ordered set of environments including mechanical pulling apart

© The Author(s) 2015
P. Bross, *The Hsp60 Chaperonin*, SpringerBriefs in Molecular Science,
DOI 10.1007/978-3-319-26088-4_1

misfolded substructures, change between hydrophobic and hydrophilic environment, proteins stuck in misfolded conformations may be reactivated to enter productive folding routes. This gradient of mechanical and biophysical parameters provided by the cycle of the machine's conformational movements and closing and opening of the lid is powered by the energy of ATP molecules that are simultaneously bound in each of the large ring subunits, hydrolysed and released. Evolution that has chosen polypeptide chains as the work horses for cellular processes has co-evolved these machines, themselves proteins, to overcome the problem that some cellular proteins are difficult to fold due to special structural peculiarities. This is especially so in the crowded and multifaceted intracellular environment where proteins undergo folding, that poses all kinds of challenges that may lead folding of peptides into a wrong direction. After a short introduction to protein folding and quality control, I will in the following focus on the mammalian chaperonin system. This will include using the broad knowledge on bacterial and eukaryotic members of the chaperonin family that can be used to illustrate and contrast the properties of the mammalian homolog.

1.2 Protein Folding and Protein Quality Control

As a basis for the understanding of the role of the Hsp60/Hsp10 chaperonin function in protein folding, here I present a short sketch of the current concepts of protein folding in living cells and its pitfalls. Basically, all the information defining the three-dimensional structure of proteins lies in the primary structure (Anfinsen 1973). Many proteins undergo posttranslational modifications and proteolytic processing that may be necessary to fold their native functional structure. However, also in these cases, the modification and processing sites are also determined by the primary structure and the processing machineries are governed by the signals lying within the amino acids sequence. The intrinsic folding code allows proteins to acquire their structure in biological environments. However, the efficiency and pace with which this is accomplished may vary widely. The current concept of protein folding is best visualized by the so-called folding landscapes (Dill and Chan 1997). Folding landscapes show, for a given polypeptide, the ensemble of possible conformations versus the free energy (Fig. 1.1). For single molecules of a given protein, there are thus many paths toward the native conformation. The existence of local energetic minima implies that when a protein folds through such pathways, this may result in a dwelling time, as it requires going back to a slightly higher energy level to leave these 'bumps' in the energetic landscape in order to reach the way toward the native conformation (Clark 2004). Intrinsic to proteins is also a tendency to aggregate (Vendruscolo et al. 2011). This is due to the fact that secondary structure elements may interact with secondary structure elements of other protein molecules of the same kind or of different kind and this may recruit more molecules finally resulting in protein aggregates. Such aggregates may even be

Chaperone blocks intermolecular interactions

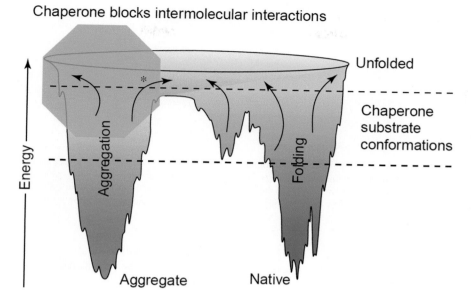

Fig. 1.1 Folding landscape and effect of molecular chaperones. The folding landscape represents the many different conformations a given protein can adapt in relation to the free energy (vertical axis). On the way to the native state, a protein molecule can follow many different pathways and this may include local energy minima in the landscape that trap the protein and require some unfolding to get back on track. Molecular chaperones interact with partially folded and misfolded proteins releasing them from energetic traps and protecting them from interactions with other molecules that might cause irreversible aggregation. See also text. Reprinted from: Clark (2004) with permission from Elsevier

energetically more stable than the native structures and aggregation is a key problem in many neurodegenerative diseases.

To control this problem caused by the aggregation propensity and liability of folding process to disturbances, both of which are intrinsic to proteins, cells have evolved helper functions that safeguard proteins on their way and sort out proteins that are on the way to aggregation. This is accomplished by molecular chaperones and proteases, which together with regulatory factors accompany proteins undergoing folding. Together, these elements form protein quality control systems (Gottesman et al. 1997) or proteostasis networks (Balch et al. 2008)—two ways of depicting such mechanisms. Molecular chaperones interact intermittently with proteins undergoing folding-typically by binding and thus shielding solution-exposed hydrophobic stretches-and keep them from unwanted interactions that might result in misfolding and aggregation (Fig. 1.2). On the other hand, special proteases degrade unfolded and misfolded proteins thus removing proteins with a low folding propensity. In a given situation, the levels and activities of these two parts of the protein quality control system determine the probability for a protein undergoing folding (and thus exposing unfolded or misfolded sections) to

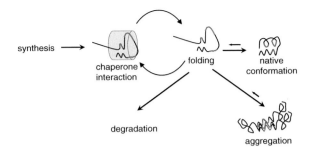

Fig. 1.2 Protein folding by interaction with molecular chaperones. Molecular chaperones interact intermittently with proteins undergoing folding-typically by binding and thus shielding solution-exposed hydrophobic stretches indicated in *red*. Thus, they keep them from unwished interactions that might result in misfolding and aggregation or premature degradation by protein quality control proteases

be safeguarded to attempt folding, or to be degraded. The balance and the overall capacity of this system are adapted by regulatory circuits-the so-called unfolded protein response systems to secure the functional proteome of cells (Haynes and Ron 2010).

Chapter 2
Historical Sketch of the Discovery and Recognition of the Function of Chaperonins

Abstract The history of chaperonin research goes back to the 1970s when the first representatives of these proteins have been described, in the first place with no clues as to their general molecular activity. A series of breakthrough studies have paved the way to our current understanding of chapeornin biology.

The history of chaperonin research goes back to the 1970s when the first representatives of these proteins have been described in the first place with no clues as to their general molecular activity-assisting other proteins in folding. A time line with the major discoveries and findings regarding chaperonin research is given in Table 2.1. The *Escherichia coli groEL and groES* genes encoding the GroEL and GroES proteins made the start in the early 1970s. The GroEL and GroES proteins encoding the large and small subunits of the chaperonin complex were found to be essential for the growth of bacteriophage T4 by sophisticated genetic screens (Georgopoulos et al. 1972; Takano and Kakefuda 1972). Electron microscopy and biochemistry experiments established that *E. coli* GroEL formed complexes consisting of two stacked seven-meric rings (Fig. 2.1; Hendrix 1979).

The RuBisCO-binding protein of plant chloroplasts saw the light of scientific publication in the early 1980s and in the end of the 1980s/beginning of the 1990s. The RuBisCO-binding protein was found to be a protein associated with the large subunit of RuBisCO (ribulose 1,5-bisphosphate carboxylase/oxygenase), a highly abundant protein in plant leaf cells. RuBisCO is the photosynthetic CO_2 fixing enzyme. The RuBisCO large subunit is synthesized inside chloroplasts whereas the small subunit is synthesized in the cytosol and imported posttranslationally. These first reports were unaware of the general function of chaperonin proteins in all cellular systems and more insight into the structural properties and mechanistic principles for the functioning of these proteins was first emerging after years of research. In the end of the 1980s, the availability of DNA sequencing technology showed that the plant chloroplast and *E. coli* proteins are evolutionarily related (Hemmingsen et al. 1988). Using an antibody directed against a 58 kDa protein from the ciliate protozoan *Tetrahymena thermophile* showed that it cross-reacted with proteins of approximately 60 kDa mass present in the mitochondria of a range of organisms such as from yeast to humans (McMullin and Hallberg 1988)

© The Author(s) 2015
P. Bross, *The Hsp60 Chaperonin*, SpringerBriefs in Molecular Science,
DOI 10.1007/978-3-319-26088-4_2

Table 2.1 Timeline of major discoveries and findings regarding chaperonin research

Year	Finding	Reference
1972	Mutations in *E. coli* abolishing propagation of λ and T4 phage later turning out to map to the *groE* locus	Georgopoulos et al. (1972); Takano and Kakefuda (1972)
1977	Indications for a large protein complex binding the large subunit of RuBisCO	Ellis (1977)
1979	The products of the *E. coli* groE operon encodes a complex composed of two stacked rings of seven-mer subunits	Hendrix (1979)
1988	Nucleotide sequences of *E. coli* GroEL and GroES and plant RuBisCO-binding protein	Hemmingsen et al. (1988)
1989	Folding of a model protein imported into yeast mitochondria involves interaction with the mitochondrial Hsp60 chaperonin and is ATP dependent	Ostermann et al. (1989)
1989	GroEL and GroES promote assembly of heterologous ribulosebisphosphate carboxylase	Goloubinoff et al. (1989)
1989	The GroES and GroEL proteins are essential for the growth of *E. coli* cells	Fayet et al. (1989)
1992	Protein folding in the cell; *seminal review on molecular chaperones and folding catalysts*	Gething and Sambrook (1992)
1993	Model for the GroEL/GroES reaction cycle	Martin et al. (1993)
1994	Crystal structure of the symmetric GroEL complex	Braig et al. (1994)
1997	Crystal structure of the asymmetric GroEL/GroES complex with bound nucleotides	Xu et al. (1997)
2002	Neurodegenerative disease caused by mutation in the gene encoding Hsp60	Hansen et al. (2002)
2005	Substrate spectrum of the *E. coli* GroEL/GroES complex	Kerner et al. (2005)
2010	Hsp60 knockout mice	Christensen et al. (2010)
2015	Crystal structure of the human Hsp60/Hsp10 complex	Nisemblat et al. (2015)

establishing the importance of this protein family in all organisms. Around at the same time, this was further emphasized by observations that knocking out the GroEL gene in *E. coli* (Fayet et al. 1989) or growing yeast cells with temperature-sensitive mutants of the chaperonin genes were not compatible with cell viability (Cheng et al. 1989). The fact that the *Tetrahymena thermophiles* representative of the family increased in expression by approximately 2–3 fold (McMullin and Hallberg 1987) together with the approximate size of 60 kDa has coined the nomenclature of Hsp60 for the mitochondrial chaperonins in all eukaryotic organisms.

As a further addition to the family, a protein found in the eukaryotic cytosol and assembling to a structure termed the TCP-1 ring complex (TRiC) was found to be evolutionarily distantly related to the Hsp60/Hsp10 and GroEL/GroES chaperonins.

Fig. 2.1 Electron microscopic pictures of purified GroEL. Negatively stained GroEL complexes seen in top view or side view (*bottom panel*). Reprinted from: Hendrix (1979) with permission from Elsevier

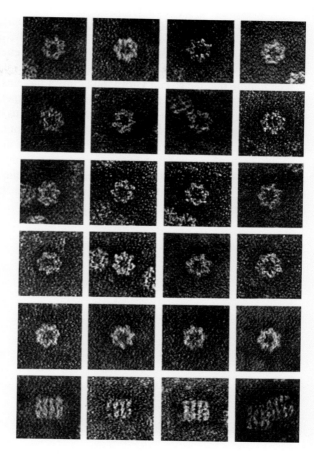

The eukaryotic TRiC chaperonin is evolutionarily rooted from Archaea, a domain of prokaryotes, whose chaperonin is termed thermosome (Trent et al. 1991). The two groups, bacterial GroEL/GroES, mitochondrial Hsp60/Hsp10, and chloroplast Cpn60/Cpn10 on one side and TRiC and thermosome on the other were then categorized as type I and type II chaperonins, respectively (Hemmingsen 1992).

In the late 1980s, more and more observations suggested that ATP-hydrolyzing heat shock proteins are involved in folding of proteins in the cell (Pelham 1986). Anfinsen's pioneering work had shown that for some proteins, folding is solely determined by the amino acid sequence. It had long been known that for some proteins, two types of helper enzymes can promote folding to the native state: (i) protein disulfide isomerases that assist in the formation of the correct disulfide bond formation, and (ii) peptidyl-prolyl-cis-trans-isomerases that mediate fast switching of the covalent bond fixing the *cis* or *trans* form of proline, which otherwise limits the mobility of polypeptide chains undergoing folding. However, it was now emerging that in vivo, the self-assembly of protein structures was facilitated by folding helper proteins that were subsequently termed molecular chaperones (Ellis and Hemmingsen 1989). Molecular chaperones bind and interact with proteins undergoing folding but are not part of the final native structure. The exact mechanisms describing how the chaperone–client interaction promoted folding were on those days still largely elusive.

The finding that the *E. coli* GroEL and GroES proteins together with ATP could reconstitute active dimeric RuBisCO in an in vitro experiment (Goloubinoff et al. 1989b) was a milestone that suggested that these proteins had a very general role for cellular function. In the ensuing years, this triggered an avalanche of experimental investigations exploring chaperonin structure, function, and molecular mechanisms. Once the basic concept of a chaperone system that formed peculiar ring complexes and was able to promote folding and assembly of proteins had been established, the door was wide open for addressing the structures, mechanisms, and functions of chaperonin complexes.

Chapter 3
Molecular Structure of Chaperonins

Abstract Experimentally determined structures of type I and type II chaperonins have shown very similar gross architecture with 'barrels' that can be closed by a lid mechanism enclosing substrate proteins. This has shed light on how the chaperonin complexes work.

3.1 Structure of Type I Chaperonins

The first detailed three-dimensional structure of a type I chaperonin was that of the *Escherichia coli* GroEL double-ring complex determined by X-ray crystallography, which was published in 1994 (Braig et al. 1994). In 1997 followed the structure of the asymmetric GroEL–GroES complex with seven bound ADP molecules in the closed ring (Xu et al. 1997). A series of additional structures have since then been determined and the coordinates of structure determinations by X-ray diffraction or electron microscopy have been deposited in the PDB databank (http://www.rcsb.org/pdb/home/home.do). For in depth reviews, see (Saibil et al. 2013; Clare et al. 2012; Clare et al. 2009; Xu and Sigler 1998).

Figure 3.1 shows representations of the GroEL double-ring complex, the asymmetric double-ring GroEL/single ring GroES complex, and the symmetric double-ring GroEL/double-ring GroES complex. The latter complexes were designated as "bullets" and "footballs," respectively, with reference to their overall shapes. The existence of football complexes in vivo and their participation in the chaperonin reaction cycle have been controversial, but the recent crystal structure of such "footballs" (Fei et al. 2014) as well as the experimental evidence supporting their role in the chaperonin folding cycle (Yang et al. 2013; Ye and Lorimer 2013) argue for a mechanistic relevance of this form of the complex.

In a representation of the asymmetric complex shown in Fig. 3.1, the front half of the complex is sliced off revealing the surface of the inner cavity. Comparing the surfaces of the upper (closed) and lower (open) chambers shows that the surface charge changes dramatically from a largely hydrophobic (gray) surface in the open conformation to a largely charged (blue for positive and red for negative) surface. This is the "folding chamber" that allows proteins to undergo folding undisturbed

© The Author(s) 2015
P. Bross, *The Hsp60 Chaperonin*, SpringerBriefs in Molecular Science,
DOI 10.1007/978-3-319-26088-4_3

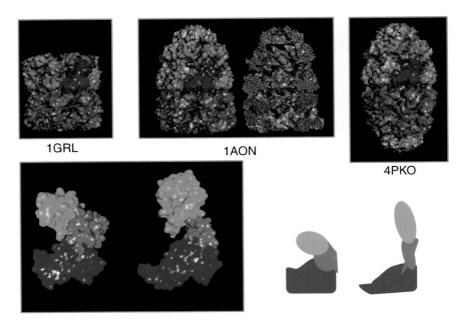

Fig. 3.1 Crystal structures of *E. coli* GroEL and GroEL/GroES complexes. The symmetric GroEL 14-mer (PDB coordinates 1GRL), the asymmetric GroEL-14/GroES complex (PDB coordinates 1AON), and the symmetric GroEL-14/GroES-14 (PDB coordinates 4PKO) are shown by surface representation. The *right* representation of 1AON shows the complex sliced open to reveal the inner cavity of the complex with *red* and *blue* indicating the negative and positive charges, respectively, and *gray* the hydrophobic surface. For one of the subunits in the ring, the equatorial (*blue*), intermediate (*red*), and apical (*green*) domains are highlighted. Enlarged representations of such subunits in the open and closed conformations, respectively, are shown on the *lower left* and a schematic cartoon is given on the *lower right*. Representations were produced using Accelrys Discovery Studio Client 4.0

by other proteins or macromolecules. Comparison of the structures with and without the bound GroES rings also reveals rather extensive conformational changes in the GroEL subunits. Binding of the GroES ring significantly enlarges the inner cavity. The structure of the GroEL subunit has been divided into three domains: the equatorial domain harboring the nucleotide binding site shown in one subunit each in blue, an intermediate domain shown in red, and the apical domain that mediates initial binding of protein substrates as well as the interaction with the GroES ring shown in green. The intermediate domain functions as a kind of hinge between the equatorial and apical domain allowing movements to be propagated between these domains. One trigger of these movements is the status of the nucleotide binding site without bound nucleotide, with ATP bound, or with ADP bound.

3.2   Type II Chaperonins: The Archaeon Thermosome and the Eukaryotic TRiC Chaperonin

The experimentally determined structures of type II chaperonins are intriguingly reminiscent of the basic architecture of the GroEL/GroES complex: double rings stacked back-to-back enclosing an inner cavity where proteins undergoing folding can be encapsulated and protected from disturbances of the folding process (see comparative schematic representations in Fig. 3.2). Also for type II chaperonins,

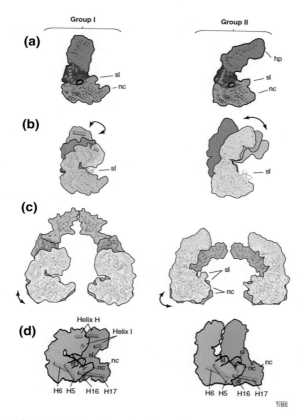

Fig. 3.2 Conformational flexibility of type I and type II chaperonins. **a** Domain organization of chaperonins from group I (*left panel* from PDB 1PCQ) and group II (*right panel* from PDB 1A6E). The apical, intermediate, and equatorial domains are shown in *green, blue,* and *pink*, respectively. The sensor loop (*sl*), the N and C termini (*nc*), and the helical protrusion (*hp*; only present in group II) are also indicated. The nucleotide binding sites, constituted by residues from both equatorial and intermediate domains, are depicted with *red ovals*. **b** Superposition of the equatorial domains of different chaperonin structures. **c** Superposition of the open (*yellow*) and closed (*blue*) ring structures of chaperonins reveals that en bloc movements of the subunits are involved in closure of the complex, accompanied by adjustments of the different domains. **d** Intra-ring contact regions. Reprinted from Yebenes et al. (2011) with permission from Elsevier

conformational rearrangement and ATP binding, hydrolysis, and ADP release trigger a cycle of binding, encapsulation, and release of proteins undergoing folding. Besides many similarities, there are some principal differences. Type II chaperonins do not use a co-chaperonin as a lid for the cavity. Rather, the lid is formed by a flexible domain that forms part of the subunits. Type II chaperonin rings are typically composed of eight subunits, but rings with nine subunits occur in some organisms. In Archaea, two or three homologous genes typically encode the subunits, whereas the eukaryotic TRiC chaperonin complex is composed of eight distinct subunits that are organized in a fixed order in the ring. These subunits are encoded by eight different genes. With regard to the folding cycle it appears that the conformational rearrangements of type II chaperonins are less coordinated in and between rings compared to the GroEL/GroES complex. For more details on the structure, properties, and mechanisms of type II chaperonins see (Yebenes et al. 2011; Lopez et al. 2015).

Chapter 4
Folding by Enclosure in the Chaperonin Cavity

Abstract In vitro studies, mostly with the bacterial GroEL/GroES complex, have revealed the chaperonin folding cycle. Analogous studies with the mammalian Hsp60/Hsp10 system have shown similarities but also some differences.

Studies using purified chaperonin proteins and purified chemically denatured potential substrate proteins are advantageous as they allow to study in detail the interactions of the substrate and the kinetics of its folding, in presence and absence of the chaperonin. Such experiments have been widely used to establish the GroEL/GroES folding cycle. The GroEL/GroES chaperonin system is by far the best studied chaperonin.

4.1 In Vitro Studies of Bacterial GroEL/GroES

In 1989, the direct demonstration that both the GroEL and the GroES protein and Mg-ATP were necessary for reconstitution of an active dimeric type II RuBisCO enzyme from denatured subunits in an in vitro reaction (Goloubinoff et al. 1989a) marked the starting signal for the elucidation of the chaperonin folding mechanism. The fact that the form II RuBisCO could be reconstituted in vitro just from the denatured subunit proteins, the GroEL and GroES proteins and ATP the authors suggested that "*because chaperonins are ubiquitous, a conserved Mg-ATP-dependent mechanism exists that uses the chaperonins for folding of some other proteins.*" This classic experiment triggered the avalanche for a huge number of investigations using such an in vitro refolding approach to elucidate the molecular mechanism underlying the chaperonin function. The *E. coli* GroEL/GroES complex became the most popular chaperonin for these studies because it could be produced in large amounts, purified in a functional form and mutants could be produced and handled using the well-established genetic tricks in bacteria. Typically, the substrate proteins used in these assays were not from *E. coli* itself, i.e. heterologous proteins from other organisms. Some of the most popular ones were rhodanese, malate dehydrogenase, and citrate synthase from mammals, and RuBisCO from photosynthetic bacteria.

© The Author(s) 2015
P. Bross, *The Hsp60 Chaperonin*, SpringerBriefs in Molecular Science,
DOI 10.1007/978-3-319-26088-4_4

Fig. 4.1 Simplified version
of the type I chaperonin
folding cycle. For details see
text

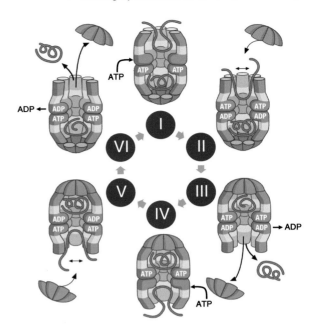

The general occurrence and essential nature of chaperonins legitimated the choice of experimentally convenient heterologous substrates to enlighten the mechanisms.

The folding of model proteins was monitored in the test tube under many different conditions. This has led to a consensus picture of the chaperonin folding cycle. A simplified version of the GroEL/GroES chaperonin folding cycle emerging from these studies is depicted in Fig. 4.1. The GroEL complex consists of two rings of each seven GroEL subunits. Different shading in the GroEL subunits distinguishes the subdomains: equatorial domain containing the nucleotide binding site and forming the interphase between the two rings, the intermediate domain that functions as a flexible link, and the apical domain (see above). The two rings are stacked back to back and are always in opposing conformations to each other so that for example the upper ring in state I is a mirror image of the lower ring in state IV. The same relationships exist between states II and V and II and VI. The subunits in the respective ring undergo coordinated conformational changes during the folding cycle. As a starting point, the apical domains of the upper ring in the figure bind the unfolded or partially folded protein mostly via hydrophobic interactions and ATP binds to the empty nucleotide binding sites in all subunits of the upper ring (I). Then a lid formed by seven GroES subunits interacts with the apical domain, binds and displaces the protein undergoing folding into the chaperonin cavity (II). These steps trigger conformational changes in the upper ring, and are accompanied by conformational rearrangements in the lower ring (III). So does binding of a GroES lid to the upper ring trigger dissociation of the GroES lid and the ADP molecules from the lower ring together with release of the encapsulated protein thus priming the lower ring for binding of another unfolded protein (IV). A timer mechanism keeps the protein undergoing folding encapsulated for a time period (estimated to approximately 10–15 s). Upon coordinated hydrolysis of the ATP molecules in the upper ring, the lower ring binds ATP molecules and concomitant binding of a GroES lid displaces the bound substrate in the lower ring (V). As for the transition of II to III with respect to the lower ring, this leads to release of the GroES lid, the substrate protein, and the ADP molecules in the upper ring (VI). Thereafter, state I is reached and the cycle can start from the beginning again.

Likely due to more tedious requirements for purification and a much less robust complex structure the mitochondrial Hsp60/Hsp10 chaperonin has so-far not experienced comparable attention by structural investigations like the GroEL/GroES system of bacteria. The mammalian mitochondrial chaperonin has the same basic arrangement of subunits in seven-meric rings of the large subunit (Hsp60) and the small subunit (Hsp10). It has, however, been heavily disputed on whether the functional chaperonin consists of stacked double rings of Hsp60, of single rings or whether both options are realized in vivo. While this issue is still not finally resolved, the recent announcement of the crystallization and x-ray structure (Nisemblat et al. 2014, 2015) insinuates that the symmetric double-ring conformation—footballs—might represent the functional unit for the mitochondrial Hsp60/Hsp10 chaperonin. Evidence for similarities and differences between the E. coli GroEL/GroES complex and mammalian type I chaperonins in mitochondria and chloroplasts has been discussed by Levy-Rimler et al. (2002). The chaperonin cycle may, in some respect, differ from the classic GroEL/GroES chaperonin cycle. However, in light of the recent resurgence of the football conformation of E. coli GroEL/GroES [(Yang et al. 2013) see above] one may expect that this question is not yet finally settled.

The particular intermediate conformations and order of events may represent important details that are pertinent to the different environments—bacterial cytoplasm and mitochondrial matrix. Nonetheless, the overall mechanism appears to be very similar. The mitochondrial Hsp60/Hsp10 complex is more dynamic with regard to dissociation of the subunits of the rings and higher conformational diversity of the conformations of the large subunit in the ring.

Chapter 5
Evolutionary Origins and Family Relations

Abstract Type I and type II chaperonins are distantly related evolutionarily and likely originate from a common ancestor. The large subunit of type I chaperonins is evolutionarily highly conserved, whereas the small subunit is less so.

5.1 A Chaperonin Ancestor

Recent analysis of the structures of the chaperonins and comparison to other protein families has led to the suggestion that a protein with a peroxiredoxin fold could have been the ancestral protein that by domain swapping evolved into a prototype chaperonin leading to type I and type II chaperonins (Dekker et al. 2011). Upon chemical modification such as oxidative modifications resulting from, for example, oxidative stress, peroxiredoxins switch between dimeric/decameric forms and high molecular weight forms. The decameric form of 2-Cys family peroxiredoxins is composed of a double ring of pentamers. Such double-ring pentamers associate to high molecular weight forms at the same time, altering its function from a peroxidase to a holding chaperone (Angelucci et al. 2013; Saccoccia et al. 2012). The stacked rings of the structural model for the high molecular weight form are reminiscent of the chaperonin ring structure enclosing an inner cavity (Fig. 5.1). Furthermore, binding of unfolded polypeptides by the high molecular weight form of peroxiredoxin appears to occur at the exposed rim of the stacked ring structure, just like for chaperonins. An early peroxiredoxin fold could thus have been the starting point for enabling evolution of protein molecules to build more complicated protein structures that need assistance during their folding process in order not to fail acquiring functional structure.

5.2 Evolutionary Relationships Between Type I, II, and III Chaperonins

Type I and type II chaperonins are evolutionarily related (Hemmingsen 1992; Trent et al. 1991). Some bacteria and a few Archaea possess both type I- and type II-like chaperonins and phylogenetic analysis has led to the suggestion that such

© The Author(s) 2015
P. Bross, *The Hsp60 Chaperonin*, SpringerBriefs in Molecular Science,
DOI 10.1007/978-3-319-26088-4_5

Fig. 5.1 Crystal structure of the high molecular weight species of *Schistosoma mansoni* peroxiredoxin. Surface representation based on PDB coordinates 3ZVJ. The double 5-mer ring is shown with alternate *blue/green* coloring to distinguish the subunits. Representations were produced using Accelrys Discovery Studio Client 4.0

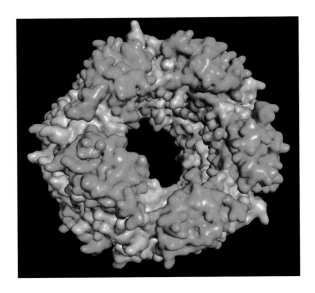

chaperonins form a specific separate third group of chaperonins termed type III (Techtmann and Robb 2010). The genes encoding these chaperonins likely originate from horizontal gene transfer events. For the archaeon *Methanosarcina mazei*, detailed interaction analyses have shown that its type I and type II chaperonins display partly overlapping but also distinct properties (Hirtreiter et al. 2009). So far, the only living organisms identified that do not possess a chaperonin are some Mycoplasma species (Wong and Houry 2004). Mycoplasma is a bacterial genus whose representatives have no cell wall and counts among the living organisms with the smallest genomes. *Mycoplasma pulmonia*, one of the species lacking a chaperonin gene, is also deficient for some other genes reported to be essential for a self-replicating minimal cell. It is therefore likely that independently living cells and organisms need a chaperonin to be able to fold at least some essential proteins whose unassisted folding would hardly take place.

Type I chaperonins have bacterial descendence and they are found in the bacterial cytosol, in the mitochondrial matrix, and in the lumen of chloroplasts. Type I chaperonin complexes consist of 7-meric ring structures of both the small and the large subunit. Two rings of the large subunit are arranged tail-to-tail and rings of the small subunit—also termed co-chaperonin—binds to the heads of the double ring thus forming lids that close the cavities of the ring shapes.

Type II chaperonins are of Archeal descendence and they are found in Archaea and in the cytosol of eukaryotic cells. Type II chaperonins form tail-to-tail stacked double-ring structures such as the type I chaperonins. They form complexes consisting of eight or nine subunits per ring but, instead of cooperating with co-chaperonins, they have a domain providing a built-in lid in the ring subunits.

5.3   High Sequence Similarity Between Representatives of the Large Type I Chaperonin Subunit

The large subunits of type I chaperonins are among the evolutionarily highest conserved protein sequences. For example, the large subunits from *Escherichia coli* and human mitochondria display almost 50 % sequence identity. This high conservation of Hsp60 proteins has proven very useful for constructing detailed bacterial species trees (Sakamoto and Ohkuma 2010).

Amino acid and structural conservation of type I chaperonins has been studied in detail by Brocchieri and Karlin (2000) and Karlin and Brocchieri (2000). Their analysis has revealed a number of characteristic patterns. Most conserved is the nucleotide binding site. Also highly conserved are hydrophobic residues that contribute to substrate binding. A cluster of intra- and intermonomer charge interactions is highly conserved presumably playing an important role for interaction with the substrate and a large number of charged residues line the central cavity in the substrate-releasing conformation. Short segments that are unaligned are generally exposed at the outside wall of the chaperonin complex.

An evolutionary tree based on the alignment of bacterial GroEL, mitochondrial Hsp60, and plastid Cpn60 sequences from eukaryotic model organisms and relevant bacterial phyla is consistent with the notion that the mitochondrial and chloroplast branches derive, respectively, from nonphotosynthetic and photosynthetic bacteria (Fig. 5.2). Photosynthetic cyanobacteria, the bacterial phylum that likely has contributed the precursor of chloroplasts type I chaperonins, typically possess two chaperonin genes whereas proteobacteria including *E. coli* and *Rickettsia prowazekii*, the obligate intracellular parasite that appears to be the closest free-living relative of mitochondria possess typically only one gene for a type I chaperonin large subunit (Lund 2009). Plant plastids usually contain several copies of the large subunit. Based on sequence conservation, these are distinguished into alpha and beta subunits and called Cpn60α and Cpn60β, respectively (Vitlin Gruber et al. 2014). Although this is not clear-cut from the evolutionary tree shown, it is likely that the two subforms derive from the two different subunits that are already present in cyanobacteria like *S. sp. PCC6803*. Alpha and beta subunits have been shown to form heterooligomers and the subunit composition of the plastid chaperonin complexes is still not well established (Vitlin Gruber et al. 2013). Plant mitochondria may possess several copies of large subunits with high sequence similarity suggesting that they have arisen by more recent gene duplications in the plant genome and originally derive from a single gene in the bacterium that was the precursor of current mitochondria. In one of the genetically best characterized plants—*Arabidopsis thaliana*—several genes encoding cpn-alpha and cpn-beta proteins are present in the nuclear genome (Suzuki et al. 2009). For one of the chloroplast large subunits, it has been proposed that it is specific for folding of one specific protein that cannot be folded by the other representatives of the large subunit (Peng et al. 2011). In contrast, mitochondria from animals and fungi typically harbor only one gene encoding the large subunit.

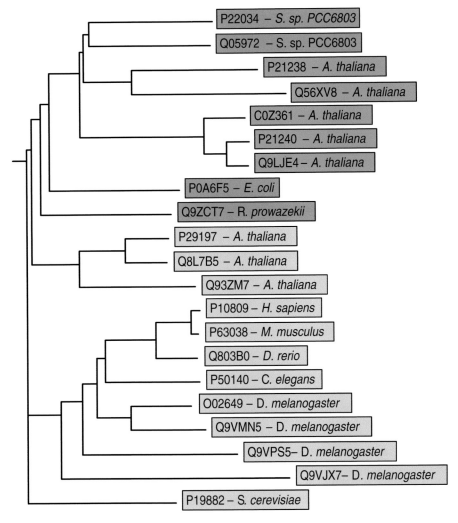

Fig. 5.2 Evolutionary tree of type I chaperonin large subunits from selected organisms. The sequences of the model organisms present in UNIPROT were aligned using CLUSTALW2 at EMBL-EBI and the evolutionary tree was plotted using the service drawgram at http://mobyle. pasteur.fr. Bacterial representatives are *shaded blue*, mitochondrial representatives *yellow*, and chloroplast representatives *green*. UNIPROT ID's and species are indicated

5.3.1 Co-chaperonin Diversity

In contrast to the large subunit, sequence conservation of the small chaperonin subunits is clearly lower as exemplified by the 32 % sequence identity between human Hsp10 and *E. coli* GroES. Most metazoans possess a unique gene encoding their mitochondrial Hsp10. Insects may possess several co-chaperonin genes, and plants have several genes for both their plastid and mitochondrial co-chaperonins (see Table 8.1). The plant plastid co-chaperonins are, in analogy to their large subunit counterparts and to distinguish them from the mitochondrial paralogs, called Cpn10's. One of the plastid co-chaperonin subunits is approximately twice the size and comprises two Cpn10-like sequences, joined head to tail, and hence termed Cpn20. The Cpn20 protein appears to form tetramers and not heptamers (Koumoto et al. 1999). How the symmetry mismatch between Cpn20 tetramers and Cpn60 heptamers is solved in the functional complexes is unclear and whether the different options with complexes formed by different versions of the large subunit together with different versions of the small subunit are related to the folding of different plastid proteins remains to be determined (Vitlin Gruber et al. 2014). For a review on chaperonin/co-chaperonin interactions, see Boshoff (2015).

An interesting finding has been made for the bacteriophages T4 and RB49, which bring their own co-chaperonins, gp31 and CocO, respectively, encoded in their viral genomes (Rao and Black 2010). In order to fold T4's major capsid protein gp23 and assemble it into proheads, the bacteriophage DNA-encoded gp31 co-chaperonin is absolutely necessary. Recombinant expression of gp31 can complement the deletion of the *groES* gene in *E. coli* (Keppel et al. 2002), while deletion of the *groES* gene otherwise is lethal. Although GroES and gp31 only display 14 % sequence identity, GroEL/gp31 complexes like GroEL/GroES complexes support citrate synthase and RuBisCO folding in vitro (Richardson et al. 1999; van der Vies et al. 1994). Structural analysis of the GroEL/gp31 complex with encapsulated gp23 has shown that this complex has a larger cavity than the classical GroEL/GroES complex allowing accommodation of the elongated gp23 molecule (Clare et al. 2009).

In this context, it should be noted that genes encoding proteins with significant sequence similarity to the large subunit of type I chaperonins that display anti-aggregation activity in vitro have been identified in bacteriophage genomes; for example, in the DNA of the *Pseudomonas aeruginosa* bacteriophage EL (Kurochkina et al. 2012). The specific characteristics of these bacteriophage GroEL homologs that appear to be closest related to bacterial GroEL proteins, has so far not been elucidated in more detail.

Artificial combination of human and *E. coli* large and small chaperonin subunits in vitro and in vivo has shown that GroEL/Hsp10 complexes are functional whereas Hsp60/GroES complexes are not (Richardson et al. 2001). Replacement of the mobile loop domain of GroES with that of Hsp10 or mutations in this domain in GroES establishes the functionality of Hsp60/GroEL-mutant complexes. The rationale for these findings was proposed to lie in the requirement of a higher affinity in the mobile loop domain of the co-chaperonin to allow functional interaction with Hsp60, compared to the interaction with GroEL. The complexity of the interaction patterns was further complicated by the finding that Hsp60 mutants could be isolated that productively function with wild type GroES both in vitro and in vivo (Parnas et al. 2012). The compensating mutations are located in the apical domain, however, not at positions expected to directly interact with the co-chaperonin.

In summary, interaction of the co-chaperonin and the large subunit is mediated by the mobile loop region in the co-chaperonin that contacts hydrophobic sites in the apical domain of the large subunit. The rest of the co-chaperonin sequence may be diverse as long as it allows the formation of a circular complex with the required size and architecture. Different forms of the architecture of different co-chaperonins may relate to modulation of specificity/preferences and capacity of different complexes to fold certain proteins.

5.4 Type II Chaperonins

Type II chaperonins comprise the thermosome of Archaea and the TRiC/CCT chaperonin of the eukaryotic cytosol. The thermosome and the CCT chaperonin are evolutionarily clearly related suggesting common descendance. Both form double-ring structures. The subunits of the thermosome of Archaea are encoded by 1–5 different genes forming double-ring complexes with 8 or 9 subunits per ring (Lund 2011). The crystal structure of the *Thermoplasma acidophilum* thermosome revealed 8-meric ring complexes with alternating alpha and beta subunits (Ditzel et al. 1998). The TRiC complex of eukaryotes is encoded by eight different genes each defining one distinct subunit of the 8-meric ring. The exact arrangement in the yeast TRiC complex has been assessed showing defined order in the rings and also fixed interring interaction architecture with two of the subunits having tail-to-tail interactions with an identical subunit in the stacked rings (Kalisman et al. 2012, 2013). A similar picture has been established by cryo-electron microscopy for the mammalian TRiC complex (Cong et al. 2010). The eight genes encoding the eight unique subunits are already present in early eukaryotes and the fact that the sequence identity between the single TRiC subunits of different eukaryotes is clearly higher than that between the different subunit genes in the same organism thus suggests that the specialization and specific localization of the subunits in the rings has happened early in the eukaryotic branch of evolution (Leitner et al. 2012).

As mentioned above, type II chaperonin complexes do not contain small subunits or co-chaperonin, rather, a domain extending the apical domain of the ring subunits can undergo conformational changes that opens and closes the cavity like a mobile lid. There are three other human genes encoding proteins with significant sequence similarity to the TRiC subunits: BBS6, BBS10, and BBS12. Mutations in these genes are associated with Bardet–Biedl syndrome (BBS), an autosomal recessive genetic disorder characterized by obesity, retinopathy, polydactyly, renal malformations, learning disabilities, and hypogenitalism as well as secondary defects including diabetes and hypertension (Zhang et al. 2013a).

Chapter 6
Chaperoning Mechanisms: Folding Helpers, Folding Protectors, or Misfolding Blockers?

Abstract Molecular chaperones facilitate folding of newly synthesized proteins and refolding of partially unfolded proteins. This occurs by a combination of various mechanisms including mechanical stretching, change of physicochemical environment, and confinement that are orchestrated and timed by the chaperonin cycle.

After the breakthrough discovery that the GroEL/GroES complex was able and necessary to refold RuBisCO in vitro, it appeared that chaperonins may catalyze folding and assembly of proteins in vivo (Cheng et al. 1990). The attractive analogy of a kind of casting machine or 3D printer that forms and determines the three-dimensional structure of the protein entering the cavity was quickly discharged by many arguments. Maybe the strongest argument against this notion was that the GroEL/GroES complex would have to have a record on or read from some unknown template the information for the many different structures it assists to form.

Molecular chaperones facilitate folding of newly synthesized proteins and refolding of partially unfolded proteins. Molecular chaperones do not confer structural information by 'molding' proteins into a specific three-dimensional shape. Rather, they protect proteins from disturbances during their folding into their three-dimensional structure that is intrinsic in their sequence of amino acids and help to overcome kinetic barriers when 'wrong' kinetically trapped folding intermediates have formed. This supports proteins undergoing folding to navigate through the folding landscape in order to reach the native structure. A number of mechanisms have been proposed through which this may be accomplished. While folded proteins typically confine hydrophobic segments from the water interface by accommodating them in the core of their native structures or by incorporating them in the lipid bilayers of membranes, unfolded proteins expose such segments during translation and while undergoing folding. Most molecular chaperones provide interacting surfaces for such hydrophobic stretches and intermittently interact with them preventing other inappropriate molecules with hydrophobic surfaces from interacting eventually initiating aggregation. Chaperonins also initially bind unfolded, partially folded, and misfolded proteins by hydrophobic interactions; however, there are a number of special properties of chaperonins that allow them to apply additional mechanisms to support folding of client proteins. In the following,

P. Bross, *The Hsp60 Chaperonin*, SpringerBriefs in Molecular Science,
DOI 10.1007/978-3-319-26088-4_6

I list some major strategies that have been suggested and that are not mutually exclusive:

(a) Confinement: Chaperonins encapsulate their client proteins for a certain time period and in this way isolate them from any interactions with other proteins.

(b) Hydrophobicity switch: Upon encapsulation the inner surface of the chaperonin cavity changes from mostly hydrophobic to mostly hydrophilic with excess of exposed negative charges (Gupta et al. 2014). Encapsulation thus both protects against inappropriate interaction with hydrophobic sections of other proteins and forces hydrophobic sections into the protein core by exposing a hydrophilic surface.

(c) Iterative annealing (Corsepius and Lorimer 2013; Stan et al. 2007; Todd et al. 1996; Kmiecik and Kolinski 2011): By repeatedly binding kinetically trapped conformers, chaperonins randomly disrupt their structure, release them in less folded states, and thus allow proteins undergoing multiple opportunities to explore pathways to their native state.

(d) Mechanical stretching: Binding of several hydrophobic segments to different chaperonin subunits of the ring to the client protein followed by the conformational movement of the chaperonin subunits with respect to each other in the chaperonin folding cycle unfolds polypeptide chains by mechanical stretching and has thus the potential of resolving unintended interactions that block further folding to the native state.

(e) Percolator hypothesis (Csermely 1999): This theoretical hypothesis proposes that chaperonins '…*by a multidirectional expansion preferentially loosen the tight, inner structure of the collapsed target protein … and during the expansion water molecules enter the hydrophobic core of the target.*' This mechanism is expected to loosen wrongly formed hydrophobic cores.

Some of these mechanisms have been tested experimentally, and others are just theoretical hypotheses. One may envision that combination of these mechanisms may be operative, or that some are more important for the folding of some proteins, and some more important for others. For an unfolded, partially folded, or misfolded protein, binding to the chaperonin by hydrophobic interaction may start a kind of roller coaster ride through different biophysical landscapes that contain all these mechanisms—mechanical stretching, change of immediate environment, confinement—that is orchestrated and timed by the chaperonin cycle. This 'ride' may then be taken once or several times depending on the nature and the status of the protein undergoing it.

6.1 Which Proteins Use and Which Proteins Need Chaperonin Assistance for Folding?

As chaperonins—with a few very special exceptions—are essential proteins in all cells and occur in two different types in different compartments like the cytosols of prokaryotes, Archaea and eukaryotes in mitochondria, and in chloroplasts, it appears that folding of certain proteins in these cellular compartments is absolutely dependent on chaperonin assistance. On the other hand, protein folding in the lumen of the endoplasmic reticulum, the mitochondrial intermembrane space, and the eubacterial periplasmic space occurs without chaperonin assistance—in spite of the fact that these compartments contain a considerable fraction of cellular proteins. A number of proteins are dually targeted with localization both in the cytosol and inside mitochondria based on alternative splicing or many other mechanisms (Yogev and Pines 2011). Such proteins have identical or almost identical primary structure and are able to fold in compartments with different types of chaperonins. In addition to that, many proteins in chaperonin containing compartments do not need assistance by these folding helpers but fold spontaneously or by assistance with other chaperones. What features then render proteins hooked up on specific chaperonin interaction to accomplish correct folding?

The question whether chaperonins assist folding of a certain subset of proteins with specific structural properties related to their folding, or whether it assists most if not all proteins in the respective environment with a bias for those proteins that have more tedious folding properties is still quite open. It has been approached by theoretical calculations on how many proteins chaperonins can maximally serve, by determining the flux of newly synthesized through chaperonin-bound intermediates, by determining chaperonin interactomes in vitro and in vivo. Each of these approaches has provided very valuable information but the jury is still out on what the situation is in living cells.

There are some main drawbacks that make it difficult to study chaperone substrates and especially to judge the degree of their dependence on the chaperone. This holds true for chaperones in general. A major challenge is the task to measure the folding propensity and yield of potential substrates. Many experiments have relied on the distinction between soluble and insoluble forms of the proteins in question equalizing folded conformation, with the protein being soluble and unfolded/misfolded conformation with being insoluble. This allows using centrifugation of the sample followed by analysis of the pellet and supernatant fractions by simple SDS-PAGE often combined with western blotting if the substrate protein in question has to be detected in complex mixtures. If functional activity can be measured, this may in most cases be the gold standard to judge acquisition of the native structure. For homo and heteroligomeric proteins a measure for judging successful folding is assessing the amounts of the oligomer.

In vivo one may also expect that misfolded and unfolded proteins are rapidly degraded by protein quality control proteases and the increase of the relative amounts of a given protein when comparing chaperonin, absent/present situations would relate to the stabilization by folding contributed by the chaperonin. There are a few other criteria that have been used, for example measuring aggregation in in vitro experiments by light scattering and studying folding assistance by the chaperonins on the basis of inhibition of aggregation.

In vivo settings face the challenge that only a small fraction of the proteins present at a certain time point undergo folding making it difficult to analytically focus on this fraction in overwhelming background of folded proteins. In addition, proteins that undergo chaperone-assisted folding cycle between chaperone-bound

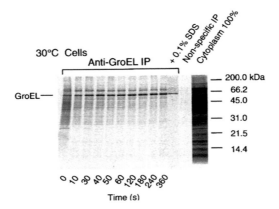

Fig. 6.1 Transit of endogenous substrate proteins through GroEL. *E. coli* (LMG194) cells were pulse-chase labeled followed by isolation of GroEL–polypeptide complexes by immune precipitation. **a** 16 % SDS-PAGE of GroEL and co-immunoprecipitated polypeptides isolated from *E. coli* labeled at 30 °C. Excess nonradioactive methionine was added at time 0 after a 15 s pulse with [35S] methionine. For comparison, an equivalent portion (100 %) of total radiolabeled cytoplasmic proteins was also analyzed. Reprinted from Ewalt et al. (1997) with permission from Elsevier

and chaperone-free situations and it is sometimes difficult to freeze labile chaperone-substrate complexes in an experimental setting.

Interactions of proteins undergoing folding are also notoriously dependent on the cellular situation. For example, 10–15 % of cytosolic proteins are in *Escherichia coli* interacting with the GroEL/GroES complex under normal conditions (Fig. 6.1) and 30 % or more of the proteins upon exposure to heat shock (Ewalt et al. 1997). On top of that, single amino acid replacements often alter the interaction (Zahn et al. 1996) and may render the protein more or less chaperonin-dependent—the differences being sometimes quite significant. In the following, I will give examples of the various approaches and summarize and discuss the intricacies and the limitations associated with the respective strategies.

6.1.1   Theoretical Considerations

Theoretical calculations have been performed based on the knowledge on the molecular composition of the *E. coli* cell and the properties of the GroEL/GroES complex. According to Lorimer (Lorimer 1996), an *E. coli* cell contains 1580 GroEL 14-mer complexes and 3072 GroES 7-mer rings. Taking into account the average protein synthesis rate, and the number and in vitro determined folding rates of GroEL/GroES complexes, Lorimer ends with 2–5 % of newly synthesized *E. coli* proteins that might find an empty GroEL/GroES complex for assisting its folding. This can be extrapolated to maximally 74 different proteins that take up all of the folding capacity of this complex. In their calculations based on similar average values for *E. coli* cells and GroEL/GroES complex properties, Ellis and Hartl (1996) find that about 7.5 % of all of the polypeptides folding in the cytoplasm of the bacterium could be folded by the GroEL/GroES complex.

The calculations are based on a number of assumptions. The most critical one is the one claiming that the kinetics of the chaperonin folding cycle measured in vitro with specific model substrates are transferable to the situation in living cells and are valid for the whole organismal proteome. They must therefore be taken with great caution, but they provide valuable estimates of what the limitations for chaperonin-assisted folding are in terms of stoichiometry. These calculations strongly argue for the notion that folding of the majority of proteins in the *E. coli* cytoplasm occurs without interaction with the GroEL/GroES complex.

6.1.2   Experimental Assessment of Proteins Interacting with the GroEL/GroES Complex

Experimental approaches have in different ways tried to assess which proteins interact with chaperonins and how large that percentage is. One of the first experiments with this aim was described by Viitanen et al. (1992). These authors investigated the binding of proteins from a total soluble protein extract prepared from *E. coli* cells to purified GroEL. By comparing the fraction of proteins binding with and without prior denaturation this experiment showed that a large proportion of *E. coli* proteins—roughly estimated to about 50 %—bound to GroEL complexes after being chemically denatured, whereas only a very small amount of non-denatured *E. coli* proteins did so. This finding is consistent with the later obtained evidence showing that GroEL rings have binding sites for hydrophobic stretches of unfolded proteins. While showing that about half of the soluble *E. coli* proteins can be bound by GroEL complexes, this does not necessarily mean that this is the case in in vivo. The theoretical calculations taking into account how many GroEL complexes are available (see above) suggested that there are by far not enough chaperonin complexes present to satisfy all these less-potential binding partners during their folding or refolding.

Using an *E. coli* strain carrying a plasmid-encoded temperature-sensitive GroEL variant gene and knocked out for the chromosomal wild-type GroEL gene, Chapman et al. (2006) cataloged the proteins that were aggregating when activity of the GroEL mutant variant was instantaneously shut off by temperature shift. These experiments suggested '*that GroEL function supports the proper folding of a majority of newly translated polypeptides.*' A series of validating experiments including pulse-chase approaches and identification of GroEL substrates well known from in vitro studies were performed to support this notion. The authors, however, note the possibility that '*aggregation of "stringent" GroEL/GroES-dependent substrates may secondarily produce an "avalanche" of aggregation ….*'

The Hartl laboratory has performed a series of studies to catalog GroEL substrates, assess whether they have special properties, and to grade their dependence on interaction with this chaperonin complex for folding. In the first study, the flux of newly synthesized *E. coli* proteins through the GroEL/GroES complex was investigated by pulse-chase/immunoprecipitation experiments (Ewalt et al. 1997). These experiments showed that 10–15 % of the newly synthesized cytoplasmic proteins intermittently bind to chaperonin complexes. This percentage increased to 30 % under heat stress, a treatment that both poses a higher burden to protein folding and results in upregulation of expression of the GroEL and GroES proteins. Bound proteins were typically released after 10–30 s. This study also suggested that the proteins interacting with the GroEL/GroES complex can be divided into three groups: one that folds without chaperonin assistance, one that also can fold independent of chaperonin help, but interact with the GroEL/GroES complex in vivo, and one group that is highly dependent on GroEL/GroES interaction for accomplishing productive folding. A follow-up study showed that approximately 300 newly synthesized proteins strongly interact with the GroEL/GroES complex one-third of which repeatedly return for refolding (Houry et al. 1999). In this study, a series of interacting proteins were identified and in a subsequent immunoprecipitation/mass spectrometry study (Kerner et al. 2005) the proteins interacting to different degrees with the GroEL/GroES complex were cataloged in detail. This study identified approximately 250 interactors, with 84 proteins being obligate GroEL/GroES substrates that will not fold to a significant degree without chaperonin interaction. Stratification of the approximately 84 obligate substrates by structural properties revealed an apparent enrichment for Tim barrel proteins. For some of the proteins the GroEL/GroES dependence was experimentally tested in the same study.

Fujiwara et al. (2010) used an in vivo system allowing to deplete *E. coli* bacteria for the GroEL and GroES proteins and performed large-scale mass spectrometry (MS) analysis to identify proteins that aggregated or were degraded due to chaperonin depletion. This approach was based on the rationale that known obligate GroEL/GroES substrates display either increased aggregation or degradation in this cell system and that their abundance in the soluble fraction is therefore observed as being significantly decreased. The picture obtained by these authors resembled to a large degree the study by Kerner et al., however, it also showed some discrepancies. So did 44 % of the obligate GroEL/GroES substrate proteins (43 of the 84 proteins classified by Kerner et al. not show a change in amounts upon GroEL/GroES depletion. On the other hand, some of the proteins earlier classified as only partially chaperonin-dependent (24 of the 75 detected by MS) and even a few by Kerner et al. classified as chaperonin-independent proteins (3 of the 38 detected by MS) displayed significantly decreased level upon chaperonin depletion. The Taguchi laboratory has performed a number of additional studies using in vitro transcription and translation systems in the presence and absence of one or several chaperone systems (Niwa et al. 2009, 2012) to globally assess the chaperone and chaperonin dependence of the *E. coli* proteome. Altogether, these studies have contributed to establish a common picture, which in broad lines shows that a small fraction of proteins in the *E. coli* cytosol is unable to fold without GroEL/GroES assistance.

Many other proteins can potentially also fold in the absence of this system but the yield may increase in its presence. The categorization of proteins into classes is somewhat blurred and environmental conditions and other factors may change affiliation of a given protein to a given class.

6.1.3   Do Specific Fold Types Have Increased Chaperonin Requirements?

A considerable fraction of the proteins classified as obligate GroEL/GroES substrates by the studies performed in the Hartl and Taguchi laboratories shares the $(\beta\alpha)8$ triose-phosphate isomerase (TIM)-barrel fold, a domain topology characterized by many long-range interactions, i.e., interactions between amino acid residues that are distant in the primary sequence. Folding of such complex structures is expected to be easily disturbed by nonnative interactions resulting in trapped folding intermediates. A study by Georgescauld et al. (2014) further explored folding of proteins with this domain structure by comparing spontaneous and GroEL-/GroES-dependent folding of the TIM barrel protein NanA from *E. coli* (EcNanA) and a structural homolog from the bacterium *Mycoplasma synoviae* (MsNanA), one of the mycoplasma species that does not have a GroEL/GroES system. These in vitro folding studies monitored by regain of enzyme activity of the denatured protein indicated that refolding of EcNanA to the native state only occurred in the presence of the GroEL/GroES chaperonin system and ATP, whereas MsNanA refolded spontaneously with very similar kinetics. Studies with another *E. coli* TIM barrel protein, DapA, in the presence or absence of the GroEL/GroES chaperonin system and ATP showed that DapA could refold without chaperonin assistance, and refolding was greatly enhanced (30-fold) in the presence of the chaperonin system. Further detailed experiments using among other methods deuterium exchange MS indicated that spontaneous refolding of EcDapA occurred by simultaneous structure acquisition of all TIM barrel elements resulting in slow folding, whereas chaperonin-assisted refolding occurred through 'segmental' structure formation. Intriguingly, spontaneous refolding of the structural homolog of DapA from the bacterium lacking a chaperonin system proceeded also fast by segmental structure formation. These data led the authors to propose '*that the chaperonin cage acts as a powerful folding catalyst for a set of proteins.*' On the other hand, this also illustrates that proteins with very similar domain structure may differ tremendously in their dependence on chaperonin assistance and the need of this folding catalysis to acquire the native structure in a biologically relevant timescale.

6.1.4   Size Limits of Client Proteins

The volume of the cavity of the GroEL ring capped by GroES in the crystal structure is 175,000 A^3 (Xu et al. 1997) and this could theoretically accommodate a globular protein of approximately 142 kDa assuming maximum packing of atoms. Experimental studies of the capacity of the *E. coli* GroEL/GroES cavity indicate that it can accommodate proteins with a maximum size of around 60 kD (Sakikawa et al. 1999). Larger proteins, for example the 82 kDa aconitase, have been shown to interact and be assisted in folding (Chaudhuri et al. 2001). Also, encapsulation by the *E. coli* GroEL/GroES complex of a dimer of the alpha and beta subunits of human branched chain ketoacid dehydrogenase (BCKDH) with a size of 86 kDa has been observed by cryo electron microscopy (Chen et al. 2006). The size of the chamber encapsulating this dimer appeared enlarged by approximately 80 % in relation to the cavity observed by X-ray crystallography. Experimental studies suggested that binding to an open ring without encapsulation of large proteins like aconitase can also promote folding to the native state (Farr et al. 2003). Yeast aconitase was shown to become 100 % insoluble in yeast strains with a temperature-sensitive Hsp60 gene, and also to a large degree (80 %) in a strain with a temperature-sensitive Hsp10 gene (Dubaquie et al. 1998). The effect of the Hsp10 ts mutation may be due to the failure of Hsp60 to undergo the normal chaperonin folding cycle when functional Hsp10 is lacking resulting in blocking of binding sites in Hsp60 rings by accumulating substrate proteins that cannot fold without encapsulation.

Taken together with the fact that bacteriophages like T4 bring their own tailored co-chaperonin to encapsulate awkwardly shaped proteins in the cavity [(Rao and Black 2010) see above] suggests that for encapsulation in the cavity there is a size limit of about 60 kDa, but that folding of some larger proteins may benefit from partial encapsulation.

Chapter 7
Sequence Variations in Proteins Affecting Chaperonin Dependence

Abstract Chaperonins assist the folding of many proteins. Some of these proteins depend more than others on chaperonin assistance and this dependence may be shifted by environmental conditions or single mutations in substrate proteins. Chaperonins facilitate evolution by allowing stepwise acquisition of mutations.

From the investigation of mutant variants, it has been indicated that single amino acid replacements can affect folding propensity and increase the requirement of proteins for chaperonin help for folding to the native state. It had been shown already in 1989 that overexpression of the GroEL/GoES proteins in bacteria (*Salmonella typhimurium*) could rescue/suppress point mutations generated by hydroxylamine mutagenesis in the genes of several diverse proteins (Van Dyk et al. 1989). The notion that increasing the availability of chaperonins by overexpression rescues mutant proteins with decreased folding propensity is supported by the experiments in which *E. coli* cells were treated with aminoglycosides. This treatment causes misreading during translation and thus random incorporation of mutations into proteins. The detrimental effects of such multiple random mutagenesis could be buffered by GroES/GroEL overexpression (Goltermann et al. 2013). Interestingly, overexpression of the bacterial Hsp70 system did also protect viability but did not promote growth in the same way as GroEL/GroES overexpression suggesting that the Hsp70 chaperone system was not in the same way able to produce functionally folded mutant proteins as the chaperonin system.

It has long been known that recombinant expression of heterologous proteins at high level in *E. coli* often results in misfolding and ensuing proteolytic degradation or accumulation of misfolded protein in aggregates (Sorensen and Mortensen 2005). Co-overexpression of molecular chaperones can in many cases rescue defective folding and allow purifying recombinant proteins in native conformation from *E. coli* cells. In particular, the GroEL/GroES chaperonin system has proven to be especially successful to counter misfolding in such settings (Baneyx 1999).

The observation that high-level recombinant expression of heterologous proteins in their native conformation in *E. coli* can be supported by chaperonin co-overexpression reiterates the notion that availability of chaperonin capacity is an important factor and that different proteins display different degrees of dependence.

© The Author(s) 2015
P. Bross, *The Hsp60 Chaperonin*, SpringerBriefs in Molecular Science,
DOI 10.1007/978-3-319-26088-4_7

This is also supported by the fact that the fraction of newly synthesized proteins interacting with *E. coli* GroEL increases upon inconvenient physicochemical conditions: exposing *E. coli* cells to high temperature (heat stress) results in an increase of newly synthesized proteins interacting with GroEL from 10–15 to 30 % (Ewalt et al. 1997).

7.1 Chaperonins as Capacitors of Protein Evolution

The concept that homologous proteins may be or not be type I chaperonin-dependent for folding although both their primary and tertiary structures are highly similar has been emphasized by a study where a GroEL-independent protein could be converted to an obligately GroEL-/GroES-dependent protein by introducing few mutations. Ishimoto et al. (2014) took a protein from the Mycoplasma species that lacks a type I chaperonin gene and performed a selection screen in *E. coli* to pick up random mutations that converted this protein to an obligate GroEL/GroES substrate. Independent multiple point mutations and even single mutations were able to confer this dependence suggesting that not a specific fold type is the determinant but rather small changes in the folding landscape that introduce local energy minima.

This observation gives some clues for the understanding of mechanisms of protein evolution. Genetic mutations are driving evolution, allowing adapting to changing environments, improving protein function and efficiency, creating novel functions, and developing organisms to higher complexity. Besides gene duplications, deletions of longer stretches of coding sequence, amino acid-replacing point mutations play an important role in this. To reach a better-adapted or novel function, a protein usually requires a number of amino acid replacements. As these occur one by one, molecular chaperones play an important role by 'allowing' intermediate protein variants to fold although the respective amino acid replacements may affect folding and stability and endanger the variant protein to be removed by the protein quality control system. Protein quality control systems with their important players molecular chaperones thus provide a limited playground for evolution by mutagenesis that allows a certain subtlety of replacements to occur without damaging the cellular function while devastating amino acid replacements that might cause serious decrease of cell viability and diseases in organisms are kept under control. For a more detailed discussion of protein evolution in general and the role of chaperones in this process, see (Bogumil and Dagan 2012; Studer et al. 2013).

Molecular chaperones may provide plasticity for protein evolution by buffering negative effects of spontaneous single mutations on folding and conformational stability and in this way allow accumulation of mutations that may improve, alter, or extend protein function in connection with the Hsp90 chaperone. This concept has originally been put forward and supported experimentally by studying examples from natural sources for the Hsp90 chaperone (Rutherford 2000, 2003; Rutherford and Lindquist 1998). For this chaperone, it has been established that high Hsp90 levels mask phenotypes that might be caused by naturally occurring sequence variants in, for example, fruit flies.

A similar case has been made for type I chaperonins, again mostly using the GroEL/GroES system as example. Single amino acid mutations are usually causing destabilization and may in some cases even abolish protein function. This is in a small fraction of cases due to the destruction of active sites or interaction sites where specific amino acid side chains play a crucial role. In the larger fraction of cases, it is due to impairment of folding/decrease of folding propensity of the protein and/or its conformational stability by the mutation. In many of the latter scenarios, one might expect that sufficient capacity of chaperonins can buffer the effect. A whole series of studies have addressed this.

A laboratory evolution experiment in which a phosphotriesterase was converted into an arylesterase by consecutive incorporation of mutations and selection of the best (variants with highest arylesterase activity) intermediates was performed with *E. coli* cells expressing normal or increased levels of GroEL/GroES (Wyganowski et al. 2013). This experiment revealed that artificially increased levels facilitated the evolution process by buffering mutations that destabilize folding intermediates and subsequently are compensated by secondary mutations in the next rounds of mutagenesis and selection.

Inspired by the observation that some endosymbiotic bacteria overexpress GroEL whereas their free-living relatives do not, Fares and colleagues designed a mutation accumulation experiment and probed the fitness of clones after more than 3000 generations (Fares et al. 2002). The fitness had greatly decreased due to accumulation of mutations. GroEL/GroES overexpression could to a large degree rescue the decrease in fitness again emphasizing the mutation effect buffering capacity of this chaperonin system.

In another laboratory evolution experiment, Tokuriki and Tawfik performed consecutive rounds of random in vitro mutagenesis of several enzyme genes that were consecutively transformed in parallel into fresh *E. coli* cells that expressed normal or increased levels of the GroEL/GroES system (Tokuriki and Tawfik 2009). Enzyme activity and expression levels of the mutated enzymes were measured after each passage. GroEL/GroES overexpression increased the number of accumulating mutations and allowed folding of enzyme variants carrying mutations in the protein core and/or mutations with higher destabilizing effects.

The role of the GroEL/GroES chaperonin in buffering of the effects of slightly deleterious mutations during evolution has been supported by further theoretical and experimental approaches. Codon optimality has been shown to covary with dependency on the chaperonin GroEL (Warnecke and Hurst 2010). This means that for example obligate GroEL/GroES substrates have a higher proportion of codons that are more prone to mistranslation for amino acids that have several synonymous codons than sporadic GroEL/GroES interactors. Another support comes from the finding that mycoplasma species that have lost their chaperonin genes display a lower evolutionary rate than their relatives that have kept the chaperonin genes (Williams and Fares 2010). Studies of different insect endosymbiont species revealed an apparent correlation of endosymbiont lifestyle, which is predicted to result in genetic drift and mutation accumulation, with acquisition of constitutive high expression of GroEL/GroES genes and accumulation of mutations in GroEL that affect substrate and GroES binding and movement of the apical domain (Fares et al. 2005). The handicap of accumulated mutations can thus be balanced by boosting the capacity of systems that allow mutations with slightly deleterious effects on in vivo folding propensity to acquire their functional structure. As protein structure has been found to be more conserved than protein sequence (Illergard et al. 2009), there appears consequently to exist flexibility for which different amino acid sequences can be used to form very similar structures. Amino acid replacements have thus apparently a more critical impact on the folding properties than on the actual structure. The protein folding problem of living organisms that are subject to genetic drift by arising gene variations has thus to be controlled by an adaptive rescue system. The relationship between protein quality control/proteostasis networks and evolvability of organisms and proteins has been discussed in a review on this subject (Powers and Balch 2013).

7.2   Disease-Associated Protein Variants

Given the results of the basic studies on the effects of missense mutations on the dependence on chaperonins for folding described above, it is not surprising-rather it is expected-that many disease-associated gene variations manifest their effect by decreasing the folding propensity of the affected protein. The King laboratory has in the 1980s and 1990s established the concept of temperature-sensitive folding (tsf) mutations using the bacteriophage P22 tailspike protein as a model (Mitraki et al. 1993). Such tsf mutations are defined as mutations that affect the folding of the affected protein but not the stability and function of this protein once it has acquired its native state. Secondary suppressor mutations, and the effect of temperature, were explored, and it turned out that such mutations could also be rescued by overexpression of the *E. coli* GroEL and GroES proteins (Gordon et al. 1994).

Inspired by the finding that folding of recombinantly expressed proteins in *E. coli* could be boosted and that folding to the native state of tsf mutants could be rescued by co-overexpressing the GroEL/GroES system, we explored whether a disease-associated variant of human mitochondrial medium-chain acyl-CoA dehydrogenase (MCAD), one of the enzymes catalyzing the first step in mitochondrial fatty acid oxidation, could be rescued by the same mechanism (Bross et al. 1993). The result was quite striking: rescue of significant amounts of active tetrameric mutant MCAD-p.Lys329Glu enzyme could be accomplished by increasing the levels of GroEL and GroES while no detectable amounts of active mutant MCAD were present in control cells. In bacteria grown at standard temperature (37 °C), chaperonin co-overexpression rescued a large proportion of the mutant MCAD protein from aggregation and allowed a minor fraction of the protein (5–10 %) to fold and assemble into active tetramers. More detailed investigations showed that chaperonin co-overexpression and reduction of the growth temperature were

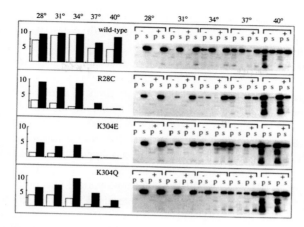

Fig. 7.1 Rescue of disease-causing mutations by chaperonin co-expression or decreased growth temperature. Solubility and MCAD enzyme activity of MCAD variants expressed *E. coli* cells at different temperatures in with (+) or without (−) co-expression of the GroES and GroEL chaperonins. *E. coli* cells were grown and induced for expression of the respective MCAD variant proteins for 1 h at the temperatures indicated. Cells were harvested, lysed, and split into soluble (s) and insoluble (p) fractions, both of which were analyzed by western blotting using anti-MCAD antibodies (*right panel*). MCAD enzyme activity was analyzed in the soluble fractions (*left panel*). This research was originally published in *J. Biol. Chem.* Bross et al. (1995)

additive (Fig. 7.1), and that at growth temperatures in the range of 28–34 °C the MCAD-p.Ly329GLu mutant protein could be rescued to approximately half the levels of wild-type MCAD expressed under the same experimental conditions (Bross et al. 1995). Notably, an artificially constructed mutant variant affecting the same amino acid position, p.Lys329Gln, folded more efficiently while still being boosted by GroEL/GroES co-overexpression. Measurements of the thermal stability of the active wild-type and mutant variants showed that the p.Ly329Glu variant displayed clearly decreased thermal stability compared to wild-type MCAD, whereas the p.Lys329Gln variant displayed thermal denaturation curves very similar to the wild-type MCAD. Lys-329 is situated in the dimer–dimer interface in an alpha helix, and it therefore appeared that the charge shift from the positive lysine to the negatively charged glutamic acid had a negative effect on assembly of MCAD subunits to the native homotetramer.

In patient's lymphoblastoid cells, the levels of the MCAD-p.Lys329Glu mutant protein could be increased by cultivating the cells at lower temperature further supporting that improving the folding conditions could partially rescue the mutation effect (Bross et al. 1995). Manifestation of MCAD deficiency in for long periods asymptomatic patients is typically triggered by fasting, but often in connection with infections and fever (Rinaldo et al. 2002). The temperature-sensitive folding mechanism may cause the levels of folded mutant MCAD enzyme to decrease at the increased body temperature. This may contribute to the precipitation of life-threatening attacks in patients, who in the absence of these stress conditions are largely asymptomatic.

This illustrated in a heterologous system that the disease-associated mutant enzyme had a higher requirement for interaction with a type I chaperonin system and that it misfolded under conditions of limiting amounts of GroEL/GroES. The relevance of this finding was validated by parallel studies in the Tanaka laboratory in which import into mitochondria, interaction with mitochondrial chaperones, and folding to the native state of the MCAD-p.Ly329Glu mutant protein were investigated using in vitro transcription/translation and import into isolated liver mitochondria from rodents (Saijo et al. 1994; Yokota et al. 1992). These studies showed that like the mutant MCAD protein, the wild-type variant was efficiently imported into the matrix space, where it sequentially interacted with first the mitochondrial Hsp70 chaperone and then the Hsp60 chaperonin. However, the wild-type protein was released from Hsp60 complexes and formed tetramers, whereas most of the mutant MCAD-p.Lys329Glu proteins remained bound to the Hsp60 complex and no tetramer was detectable. Knockdown of Hsp60 in human HEK293 cells showed that this treatment affected folding and assembly to the native tetramer for both wild-type MCAD and the MCAD-p.Lys329Glu mutant protein (Corydon et al. 2005). Taken together these complementary approaches suggest that the MCAD-p.Lys329Glu mutation renders the protein more dependent on chaperonin assistance, likely requiring many more rounds of binding and release to the chaperonin complex.

Co-overexpression of the GroEL/GroES chaperonin system was also applied for the evaluation of other disease-associated MCAD mutations (Andresen et al. 1997, 2001; O'Reilly et al. 2004) and it has recently been revitalized to characterize a broad spectrum of MCAD missense mutations detected in newborn screening programs (Koster et al. 2014; Maier et al. 2009). It has to be noted here that overexpression of the *E. coli* Hsp70 chaperone system (DnaK/DnaJ/GrpE) had a negative effect on folding to the active conformation of both wild-type and the p.Lys329Glu mutant MCAD protein (Hansen et al. 2005). This observation suggests that the Hsp70 chaperone system, when present in high amounts, negatively affects MCAD folding possibly by scavenging unfolded and partially folded MCAD molecules from interaction with the GroEL/GroES chaperonin system.

Some of the disease-associated MCAD mutant proteins cannot be rescued to significant levels, but some are partially or even fully rescued to wild-type activity levels, in the applied system and conditions. Rescue potential appears thus to be specific for particular mutations.

GroEL/GroES overexpression has been applied to study effects of disease-associated mutations in other proteins. A prominent example is phenylalanine hydroxylase (PHA), the enzyme mutated in most cases of phenylketonuria. Also for disease-associated mutant variants of this protein, co-overexpression of GroEL and GroES did result in rescuing of folding and activity (Pey et al. 2003; Martinez et al. 2008). PHA is a cytosolic protein that under normal conditions has only the cytosolic type II chaperonin system TRiC available. Whether wild-type PHA is chaperonin-dependent and whether both type I and type II chaperonins can assist its folding is currently not clear. The so-called pharmacological chaperones, which in the case of PHA are analogs of the natural cofactor tetrahydrobiopterin, have likewise been successfully used to rescue PAH mutant variants. This is consistent with the concept that such mutant variants are impaired in folding and that all measures that support, protect, or stabilize folding to the native state, be it increased chaperone help or ready availability of stabilizing cofactors, have the potential to partially or fully rescue the defect.

7.3   Studies of the Yeast Hsp60/Hsp10 System

The so-far most detailed investigations aiming to pinpoint substrates of the mitochondrial Hsp60/Hsp10 chaperonin have been performed in the yeast *Saccharomyces cerevisiæ*. These investigations have been based on the availability of yeast strains with temperature-sensitive mutations in the yeast Hsp10 or Hsp60 genes. The yeast temperature-sensitive lethal strain mif4 was widely used in these investigations. It was produced by chemical mutagenesis of a yeast strain defective for the yeast cytosolic ornithine transcarbamylase (OTC), an enzyme in the urea cycle, and carries an inducible transgene encoding human mitochondrial OTC (Cheng et al. 1989). The mif4-ts lethal mutation was selected by scoring for mutants that at the nonpermissive temperature imported the human OTC into mitochondria but failed to produce active human OTC. The mif4-ts mutation was found to be situated in the yeast Hsp60 gene and growth at the nonpermissive temperature resulted in accumulation of Hsp60 protein in an insoluble conformation. Sequencing identified a p.Gly319Asp point mutation in Hsp60 gene of the mif4 mutant strain (Dubaquie et al. 1998).

Hohfeld and Hartl (1994) selected a yeast Hsp10-ts mutant (p.Pro36Ser) that, when expressed as the sole cellular Hsp10 protein, was lethal at the nonpermissive temperature and caused reduced growth at the permissive temperature. Unlike the wild-type yeast Hsp10, the Hsp10-p.Pro36Ser mutant protein did not bind to GroEL complexes. Import of the alpha subunit of the mitochondrial matrix processing protease and the Rieske iron-sulfur protein, a subunit of complex II of the respiratory chain, resulted in accumulation of insoluble aggregates of these proteins.

In an attempt to get a broader picture of the mitochondrial chaperonin substrates, Dubaquie and coworkers employed strains with temperature-sensitive inactivating mutations in Hsp10 and Hsp60, respectively (Dubaquie et al. 1998). Solubility of the proteins produced by in vitro transcription and translation of total yeast mRNA after import into control and the respective mutant mitochondria at the nonpermissive temperature showed an increase in insoluble proteins in the mitochondria from cells carrying a temperature-sensitive mutation in Hsp60 (mif4). Identification of proteins displaying increased insolubility in the mutant strains was performed by a 2D-gel/MS approach.

7.4   Substrates of Type II Chaperonins

Compared to the GroEL/GroES chaperonin complex, less is known on which substrates use and are dependent on type II chaperonin complexes. In the following a short summary on the experimental evidence for this type of chaperonins is described.

7.4.1   The Interactome of the Archaea Thermosome

The proteins interacting with the thermosome of the archaeon *Methanosarcina mazei* have been studied experimentally (Hirtreiter et al. 2009). This archaeon possesses genes for both a thermosome and a type I GroEL/GroES complex and the study evaluated the differences in the preferences for substrate proteins. Here was partial overlap between the preferences of these two chaperonin types for interacting proteins but also distinct differences. The thermosome interactors comprised a much broader spectrum of domain folds and a greater fraction of large multidomain proteins. This may suggest a greater specialization for a certain structural class of proteins of type I compared to type II chaperonins.

7.4.2   The Interactome of the Mammalian TRiC

Actin and tubulin, highly abundant components of the cytoskeleton, account for the most predominant substrates in terms of number of molecules processed by the eukaryotic TRiC chaperonin. This has in the first phase eclipsed that many other cytosolic proteins also use interaction with TRiC during their folding process. Yam et al. (2008) applied a series of approaches to narrow down the TRiC interactome in human cells. Pulse labeling and immunoprecipitation followed by display of the immunoprecipitated proteins on 2D gels revealed that a large collection of newly synthesized proteins interacts with TRiC. After a 30-min chase period no TRiC interacting proteins were present anymore. Comparison of the proteins intermittently binding to TRiC during translation (in vivo) to the binding of a denatured labeled protein extract (in vitro) showed very different patterns with less than 10 % overlap. Interestingly, there was a very similar binding pattern in an experiment when the same denatured protein extract was exposed to GroEL or TRiC, suggesting that these two different chaperonin types have similar binding specificities in vitro.

Folding of tubulin and actin is blocked in yeast strains with mutations in TRiC subunits (Chen et al. 1994). Folding of beta-actin in vitro requires the presence of the TRiC complex and ATP. The bacterial GroEL/GroES or the mitochondrial Hsp60/Hsp10 cannot replace TRiC although they bind unfolded beta-actin (Tian et al. 1995). Actin folding can be accomplished in *E. coli* lysate supplemented with TRiC (Stemp et al. 2005) but not lysates containing the GroEL/ES chaperonin only (Tian et al. 1995). It is therefore assumed that the two different chaperonin types recognize different folding intermediates of the target protein and that this may result in different specificities and capacities to support folding of certain proteins. Two possible mechanisms may account for this: one proposing that the TRiC chaperonin works together with the yeast GimC protein (prefoldin in mammalian cells) (Siegers et al. 1999) and another stating that folding of beta-actin requires a specific mechanical domain expansion that can be performed by TRiC but not by type I chaperonins (Villebeck et al. 2007).

7.5 What Determines the Dependence of Proteins for Chaperonins?

In synthesis, the evidence presented in this section establishes that chaperonins assist the folding of many proteins, and that whereas some of them depend more than others on chaperonin assistance, this dependence may be shifted by more or less favorable environmental conditions; also, some structural domains are more prone to require chaperonin assistance and some proteins appear to fold only with the assistance of type I or type II chaperonin. On the other hand, single amino acid changes can render a given protein more or less dependent on chaperonin assistance, and variant proteins that by mutations are impaired in folding/require more chaperonin assistance, and are tolerated in cells that have increased/sufficient chaperonin levels, thus supporting evolution by stepwise mutation acquisition. Chaperonins appear therefore to be necessary to secure folding of peculiarly structured protein domains to allow genetic drift/evolution and they have in their present form coevolved to the cell and tissue type and intracellular subcompartment with their respective proteomes. Evidences for substrate preferences of the *E. coli* GroEL/GroES system have been discussed and evaluated in a review by Azia et al. (2012).

Chapter 8
Genetic Organization of Type I Chaperonin Genes

Abstract In vertebrates the genes encoding the large and small subunit of the Hsp60/Hsp10 complexes are arranged on the chromosome in a head-to head architecture with a bidirectional promoter. In bacteria, the genes for the two subunits are organized in an operon. Both organizations appear to secure expression of both subunits at a fixed ratio.

Type I chaperonins encoded by genes for the large (GroEL) and the small (GroES) subunits are found in all bacteria with only a few exceptions, namely some Mycoplasma species (Wong and Houry 2004). Bacteria typically have one gene for the large and one for the small chaperonin subunit. However, there are bacterial species with several or even multiple chaperonin and co-chaperonin genes, some of which have special functions termed as moonlighting functions (Lund 2009; Henderson et al. 2013). In *E. coli* and many other bacteria, the two genes are organized in the same operon, regulated by a single promoter.

In eukaryotes, type I chaperonins are present in organelles originating from endosymbiosis, namely mitochondria and chloroplasts. In plant chloroplasts and mitochondria, several chaperonin and co-chaperonin genes (see below) coexist, whereas most animals and fungi possess one mitochondrial Hsp60 and one Hsp10 gene. Also, insects may have several copies. The evolutionary tree with sequence relationships for type I chaperonins from humans and selected other organisms shown above (Fig. 5.2) illustrates evolutionary proximity and relationships between the multiple type I chaperonin genes.

In mammals, the genes encoding Hsp60 and Hsp10 are situated in a head-to-head arrangement with a common bidirectional promoter (Ryan et al. 1997; Hansen et al. 2003). This configuration applies also for other mammals and metazoans like *C. elegans* (Martin et al. 2002) and *Danio rerio* (Table 8.1). All these organisms possess one gene each for the two proteins. The head-to-head architecture is a rather common finding in eukaryotes and functions of such an arrangement appear to be to enhance co-regulation and to enable rapid response to different stimuli (Grzechnik et al. 2014). The fungus *S. cerevisiae* has also only one copy each for a Hsp60 and Hsp10 homolog, respectively. However, here the two genes are localized on different chromosomes thus requiring separate promoters.

© The Author(s) 2015
P. Bross, *The Hsp60 Chaperonin*, SpringerBriefs in Molecular Science,
DOI 10.1007/978-3-319-26088-4_8

Table 8.1 (A) Type I chaperonin genes in selected organisms: large subunit. (B) Type I chaperonin genes in selected organisms: co-chaperonins

Organism		Chromosome	Exons	UNIPROT ID	Chromosome location in relation to co-chaperonin gene
(A)					
H. sapiens		2	12	P10809	Head-to-head
M. musculus		1	12	P63038	Head-to-head
D. rerio		9	11	Q803B0	Head-to-head
C. elegans		III	6	P50140	Head-to-head
D. melanogaster		X	4	O02649	Unlinked
		2L	1	Q9VPS5	Unlinked
		2L	3	Q9VMN5	Unlinked
		2L	2	Q9VJX7	Unlinked
S. cerevisiae		VII	1	P19882	Unlinked
A. thaliana	Mito	5	17	Q8L7B5	Unlinked
		3	16	P29197	Unlinked
		3	17	Q93ZM7	Unlinked
	Chloro	2	8	P21238	Unlinked
		5	15	C0Z361	Unlinked
		1	15	P21240	Unlinked
		3	15	Q9LJE4	Unlinked
		5	9	Q56XV8	Unlinked
R. prowazekii		–	1	Q9ZCT7	Sequential/operon
S. sp. PCC6803		–	1	Q05972	Sequential/operon
		–	1	P22034	Unlinked
E. coli		–	1	P0A6F5	Sequential/operon

Organism		Chromosome	Exons	UNIPROT ID
(B)				
H. sapiens		2	4	P61604
M. musculus		1	4	Q64433
D. rerio		9	5	Q6IQI7
C. elegans		III	4	Q965Q1
D. melanogaster		3R	2	Q9VFN5
		3L	3	Q9VU35
S. cerevisiae		XV	1	P38910
A. thaliana	Mito	1	3	P34893
		1	3	Q8LDC9
	Chloro	?	?	O49306
		5	6	O65282
R. prowazekii		–	1	Q9ZCT6
S. sp. PCC6803		–	1	Q05971
E. coli		–	1	P0A6F9

The arthropod *D. melanogaster* has several expressed genes encoding Hsp60 and Hsp10 homologs and the chaperonin and co-chaperonin encoding genes are not localized close to each other. In the plant model organism *A. thaliana*, there are also multiple genes encoding type I chaperonin subunits and this is further complicated by the necessity to fuel both mitochondria and chloroplasts with type I chaperonins. None of the *A. thaliana* genes encoding Hsp60 homologs locate in the vicinity of a gene encoding Hsp10 homologs. The head-to-head or operon organization is likely securing that the two genes are simultaneously transcribed at a given ratio. It is however unclear how the ratio of Hsp60 to Hsp10 is regulated in *S. cerevisiae* and other organisms that do not have these closely linked architectures.

Chapter 9
Regulation of Type I Chaperonin Gene Expression

Abstract Expression of Hsp60 and Hsp10 is regulated by a combination of pro-moter elements: high constitutive expression can be modulated by binding sites for transcription factors that respond to cellular stress and endocrine stimuli.

9.1 Bacterial GroEL/GroES Expression

Transcription of the *E. coli* GroEL/GroES operon is strongly increased upon folding stress as for example generated by heat shock (Zeilstra-Ryalls et al. 1991). The increased induction upon heat shock is mediated by the sigma-32 transcription initiation factor (Guisbert et al. 2008). Transcription of sigma-32 increases upon heat shock and the stability of the translated sigma-32 polypeptide is at the same time enhanced. An immediate effect of heat shock is the sequestration of chaper-ones by unfolded and misfolded proteins, which releases an existing pool of chaperone-bound sigma-32 molecules and allows them to bind to promoter ele-ments and increase or initiate translation of heat shock-responsive genes. The GroEL/GroES operon is only one of these genes; other chaperones like the bacterial Hsp70 system and protein quality control proteases like Lon possess also sigma-32 binding sites. A similar principle appears to be present in many other bacteria as exemplified by *B. subtilis* (Schumann 1996).

9.2 Regulation of Expression of the Mammalian Type I Chaperonin Genes

Transcriptions of the *HSPD1* and *HSPE1* genes encoding Hsp60 and Hsp10, respectively, are regulated by the bidirectional promoter situated between the two genes (Fig. 9.1). Such a head-to-head arrangement of two genes is rather common in the human genome (Adachi and Lieber 2002). Coupling of two genes with

© The Author(s) 2015
P. Bross, *The Hsp60 Chaperonin*, SpringerBriefs in Molecular Science,
DOI 10.1007/978-3-319-26088-4_9

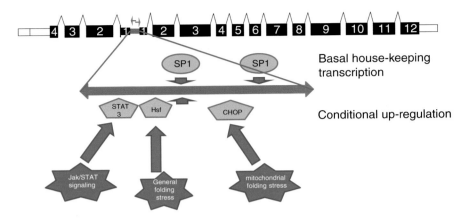

Fig. 9.1 Transcriptional regulation of expression of human Hsp60 and Hsp10 by the bidirectional promoter. The bidirectional promoter (*red*) directs transcription in both directions. Constitutive expression is mediated by SP1 elements. Transcription can be further up-regulated by binding of the following transcription factors: HSF1 to heat shock elements, STAT3 to STAT3 binding sites, and CHOP and CEPB/β to CHOP-binding sites

head-to-head arrangement may secure similar transcription rates of genes that encode interacting proteins. However, there are also many genes with this architecture where there is no such relationship. Other functions of this arrangement have been suggested to allow paused polymerases to remain poised for initiation of rapid gene expression upon stimuli and/or this may be a way to ensure the maintenance of high levels of basal gene expression (Wakano et al. 2012).

Transcription of the human *HSPD1* gene starts alternatively at one major site and two minor transcription initiation sites resulting in the inclusion of different sequences as the first exon (Hansen et al. 2003). Exons 2–12 are the same for all three transcripts. The different first exons are noncoding and the consequences for splicing and translation of the different first exons are currently not clear. It may be possible that minor transcripts are expressed during development or that translation initiation can be regulated differentially in the different forms. A number of other transcripts, all of which only comprise a fraction of the 12 exons, are reported in the ENSEMBL database ((Cunningham et al. 2015) (http://www.uniprot.org)). The origin, nature, and possible relevance of these transcripts are currently unclear.

For the *HSPE1* gene, one major transcript with four exons has been observed as well as a couple of minor transcripts with unknown relevance. The first exon of the standard *HSPE1* transcript codes just for the initiator methionine.

The bidirectional HSPD1/HSPE1 promoter (Fig. 9.1) contains two SP1 sites that likely are responsible for high-level housekeeping transcription of both genes. On top of that, a number of conditionally activated promoter elements are present that allow modulating the expression. The first discovered elements of this kind were the heat shock factor 1 (HSF1) regulated heat shock elements. HSF1 induces chaperone genes and a wide variety of genes affecting gene transcription and translation upon protein-damaging stress in the cytosol (Vihervaara and Sistonen 2014). Traditionally, HSF1 activation and ensuing transcriptional responses have been studied in cells or animals subjected to heat shock. Extreme protein structure-damaging stress like heat shock leads to silencing of overall transcription and stalling of translation and HSF1-regulated genes maintain or even upregulate their transcription and translation under these circumstances. Steady-state Hsp60 protein levels are typically increased two- to threefold in heat-shocked cells, for example in *Tetrahymena* (McMullin and Hallberg 1987).

In addition to the regulated heat shock factor-regulated induction pathway, the *HSPD1/HSPE1* promoter also contains binding sites for transcription factors involved in a protein structure damage response confined to mitochondria: the mitochondrial unfolded protein response (mtUPR) that is induced by accumulation of unfolded proteins in the mitochondrial matrix (Zhao et al. 2002). With regard to the *HSPD1/HSPE1* promoter, this induction is mediated by binding sites for the transcription factors CHOP and CEPB/β. CHOP-binding sites and induction by accumulation of misfolded proteins in the mitochondrial matrix have also been found for other mitochondrial protein quality control components. These include the ClpP protease subunit, the AAA+ protease YME1L1 of the inner membrane, the mitochondrial Hsp70 family chaperone co-chaperone mtDNAJ, the mitochondrial matrix processing protease β-subunit (MPPβ), and TIM17A, a component of the import machinery (Aldridge et al. 2007). The promoters of the encoding genes harbor a CHOP-binding site plus at least one of the two other sequence elements termed MURE1 and MURE2 (for the mitochondrial unfolded protein response element) (Aldridge et al. 2007). The *HSPD1/HSPE1* promoter lacks MURE elements. Transcription of the CHOP and CEPB/β transcription factors themselves is up-regulated upon misfolding stress in the mitochondrial matrix, however, also by misfolding stress confined to the endoplasmic reticulum. For the mtUPR it has been indicated that it requires an AP-1 site in the promoters of the genes encoding the CHOP and CEPB/β transcription factors. AP-1 sites are recognized by the c-jun transcription factor suggesting that the JNK pathway is involved in mtUPR activation. Next to the AP-1 site the CHOP and CEPB/β promoters carry ERSE sites (endoplasmic reticulum stress-responsive element). In this way, specificity for erUPR- or mtUPR-directed transcription induction of the CHOP and CEPB/β genes may be accomplished. It should be noted that the CHOP transcription factor is playing a role in apoptosis induction. Particularly, high levels of CHOP protein triggered by ER, mitochondrial or other stresses may trigger apoptosis as the ultima ratio when cells are irreparably out of balance.

The observation that Hsp60 and Hsp10 were up-regulated by interferon γ in C6 astroglioma cells has led to the identification of a binding site for the transcription factor STAT3 in the *HSPD1/HSPE1* bidirectional promoter (Kim et al. 2007). It had been shown that leptin treatment increases Hsp60 expression in the rat pancreatic acinar cell line AR42J (Bonior et al. 2006) and detailed analyses of leptin and leptin receptor-deficient mouse models and cultured hypothalamic cells revealed that the STAT3 binding site is responsible for modulation of Hsp60 expression by leptin (Kleinridders et al. 2013). Impaired leptin signaling is thus accompanied by reduced Hsp60 (and likely also Hsp10) expression in target tissues which may possibly lead to mitochondrial dysfunction caused by chaperonin deficiency. A series of experiments with mouse and cellular models suggested that hypothalamic insulin resistance as observed in type 2 diabetes patients was associated with mitochondrial dysfunction and downregulation of the Hsp60 chaperone. Importantly, type 2 diabetic patients also displayed decreased Hsp60 expression in brain suggesting a mechanistic connection between defective leptin regulation of Hsp60 expression and insulin-resistant states.

Besides the above-described promoter elements, there may possibly be further transcription regulation pathways that modulate expression of Hsp60 and Hsp10. In *C. elegans* a regulatory circuit depending on the status of the mitochondrial membrane potential has been described (Nargund et al. 2012). The worm ATFS1 protein contains both a mitochondrial targeting sequence and a nuclear localization signal. When mitochondria are well functioning, ATFS1 is imported into the mitochondrial matrix and degraded by Lon protease, apparently not fulfilling any function in the matrix. If mitochondria are dysfunctional and the membrane potential is low, ATFS1 cannot efficiently be imported anymore and then its nuclear localisation signal dominates resulting in import into the nucleus where it functions as a transcription factor. This increases, among many other things, transcription of the worm gene encoding Hsp60. A consensus sequence for ATFS1 binding has been determined and such an element is present in the worm Hsp60 promoter (Nargund et al. 2015). Human cells possess a number of homologs to the worm ATFS1 gene, but it is currently not yet clear whether a similar mechanism exists in humans.

9.3 Hsp60 and Hsp10 Dysregulation in Cancer Cells

Molecular chaperones—many of which are stress induced—have been observed up-regulated in a plethora of tumor cells and tumor stages suggesting that increases in HSP expression seem to be involved not only in tumor development, but also in the acquisition of drug-resistant phenotypes (Lianos et al. 2015; Calderwood et al. 2006). Upregulation of molecular chaperones allows tumor cells to survive in the presence of a high load of misfolding mutant proteins and to counteract apoptosis mechanisms. There are numerous reports on Hsp60 upregulation or downregulation in cancer cells (Cappello et al. 2008; Khalil et al. 2011). For example, overexpression of HSP60 associated with a poor prognosis of osteosarcoma (Mori et al. 2005) and Hsp60 has been observed up-regulated in cervical cancer in women (Hwang et al. 2009).

The transcriptional regulation of Hsp60 by STAT3, an oncogene, has been mentioned above. Increased expression of Hsp60 in cancer cells may thus be accomplished by this transcription factor. Multiple pathways including interleukines regulate JAK–STAT signaling in cancer cells thus affecting cell proliferation, survival, invasion, and immunosuppression (Yu et al. 2014). STAT3 and STAT5 have been demonstrated to be the most important STAT transcription factors for cancer progression and they are therefore attractive candidates for cancer therapy (Yu et al. 2014). The role of STAT3-mediated upregulation of Hsp60 and Hsp10 protein expressions in cancer development and progression may be brought about by safeguarding mitochondria from damages caused by accumulation of misfolded proteins in the matrix that otherwise might induce apoptosis. How central this role is for cancer development and whether treatment strategies specifically interfering with Hsp60/Hsp10 upregulation would be beneficial remain an open questions for cancer research.

Chapter 10
Subcellular Localization

Abstract Hsp60 is mostly localized in the mitochondrial matrix, but localization in other cellular subcompartments as well as secretion have been observed.

The major cellular site of localization of the Hsp60 and Hsp10 proteins is the mitochondrial matrix. Hsp60 possesses a 25-amino-acid-long mitochondrial targeting sequence, whereas Hsp10 does not have a cleaved mitochondrial targeting sequence. For the yeast Hsp10 protein import into isolated mitochondria has been demonstrated and the targeting function has been suggested to reside in a short N-terminal extension in relation to bacterial Hsp10s that may form an amphipathic alpha helix (Rospert et al. 1993).

Localization of Hsp60 to the mitochondrial matrix appears to apply for most cells and tissues studied as documented by a large body of immune fluorescence microscopy analyses as well as cell fractionation and mitochondrial subcompartmental fractionation showing localization of Hsp60 to the mitochondrial matrix and many studies have used the Hsp60 protein as a marker for this compartment. Evidence for Hsp10 is scarce, but it is assumed that it follows localization of Hsp60 as the two proteins need to interact to fulfill the chaperonin function. As both Hsp60 and Hsp10 are synthesized by ribosomes in the cytosol and imported posttranslationally, there will always be a small amount of these proteins in the cytosol—on the way to mitochondria.

There are a number of reports indicating directly or indirectly that Hsp60 is present in the cytosol. Itoh and colaborators (Itoh et al. 1995) reported purification of Hsp60 comprising the N-terminal mitochondrial targeting peptide from the cytosol of porcine liver. Later it was described that Hsp60 was present in similar amounts in both the cytosol (as precursor) and mitochondria (as the mature protein without targeting sequence) in the rat's kidney (Itoh et al. 2002). Knowlton and coworkers published a number of studies indicating non-mitochondrial localization of Hsp60 in cardiac myocytes and suggested a possible anti-apoptotic function of cytosolic Hsp60 (Gupta and Knowlton 2002; Kirchhoff et al. 2002; Lin et al. 2007). Secretion of Hsp60 by cancer cells and differential amounts of Hsp60 in the blood stream in healthy versus cancer patients have been reported (Merendino et al. 2010;

P. Bross, *The Hsp60 Chaperonin*, SpringerBriefs in Molecular Science,
DOI 10.1007/978-3-319-26088-4_10

Pockley et al. 2014). Also, a immunoregulatory role of extracellular Hsp60 as costimulator for regulatory T-cells has been indicated (Zanin-Zhorov et al. 2006). In this regard, a protein termed early pregnancy factor that was detected in mammals in maternal serum within 24 h of fertilization turned out to be identical with Hsp10. Whether the Hsp10 protein is actively secreted upon fertilization or is just a marker for necrosis is still not clear.

Most of these studies on extramitochondrial and extracellular functions and roles of Hsp60 and Hsp10 in humans have been reported by a single laboratory and further studies are necessary to underpin such functions that likely are not related to the protein folding function.

Chapter 11
Posttranslational Modifications

Abstract Many posttranslational modifications of mammalian Hsp60 and Hsp10 have been described. The function of these modifications is currently unclear.

A long list of posttranslational modifications (PTMs) of mammalian mitochondrial chaperonin proteins has been recorded based on global cellular or mitochondrial PTM investigations. This comprises a large number of lysine acetylation sites, lysine succinylation sites, lysine malonylation sites, and phosphorylation sites at serine residues. The current list can be viewed at the UNIPROT protein database (http://www.uniprot.org). In addition, a recent report stresses that chaperones represent particularly highly modified proteins (Cloutier and Coulombe 2013) and that there may exist a special PTM pattern or code that regulates chaperone activities. Furthermore, Hsp60 might be subjected to lysine methylation as it apparently interacts physically with a specific lysine methyltransferase (Cloutier et al. 2013). Another investigation argues that a specific pattern of lysine biotinylation and methionine sulfoxidation occurs in Hsp60 and hypothesizes that this pattern may have a functional role in eliminating reactive oxygen species (Li et al. 2014). These authors propose that biotin addition to certain lysines could catalyze sulfoxidation of neighboring methionines by ROS like H_2O_2, and reduction of the sulfoxidizded methionines by methionine sulfoxidases could scavenge ROS in the mitochondrial matrix. Mammalian Hsp60 possesses a carboxy-terminal sequence with glycine and methionine repeats. This sequence is present in most type I chaperonins with some exceptions, namely chloroplast Hsp60 homologs. In the crystal structure of GroEL, this part of the protein is not resolved suggesting that it is not specifically structured. Deletion of the entire C-terminal repeat region in *Escherichia coli* GroEL showed that the truncated protein was functional in vivo and that it had chaperonin activity in in vitro assays (Mclennan et al. 1993). Modification of the repeat and its length in *E. coli* GroEL revealed that these repeats play a role for the folding rate and preference od substrate size suggesting that it is important for confinement and kinetics (Tang et al. 2006).

© The Author(s) 2015
P. Bross, *The Hsp60 Chaperonin*, SpringerBriefs in Molecular Science,
DOI 10.1007/978-3-319-26088-4_11

Chapter 12
Variations in Hsp60 and Hsp10 in Humans

Abstract Only few polymorphisms including one amino acid replacement have been observed in human Hsp60 and Hsp10.

The amino acid sequences of Hsp60 and Hsp10 proteins in humans display only very few variations (Fig. 12.1). There is only one amino acid variation with an allele frequency of more than 1 % fulfilling the criterion for a genetic polymorphism: p.Gly563Ala. This variation was found with an allele frequency of 1.3 % in Danish control individuals (Bross et al. 2007) and has been detected with frequencies of 1.7 and 0.3 % in European Americans and African Americans, respectively, in the NHLBI Exome Sequencing Project (http://evs.gs.washington.edu/EVS/). This variation localizes to the Gly-Met repeats at the carboxy-terminus of the protein (see above). Heterozygosity of this variant has been suggested to be a modifying factor in hereditary spastic paraplegia patients carrying disease-causing mutations in the *SPAST* gene (Hewamadduma et al. 2008). So far, a modifying role of the p.Gly536Ala variation in patients with disease-causing *SPAST/SPG4* mutations could not be further established: in a Danish spastic paraplegia patient family carrying a *SPAST/SPG4* mutation and this variation in some of the individuals there was no correlation with phenotype severity, however, the small number of individuals did not allow more general conclusions (Svenstrup et al. 2009). Other published studies that included monitoring of the p.Gly536Ala variation in spastic paraplegia patient cohorts did not detect individuals carrying this variation (Racis et al. 2014; Luo et al. 2014) and search for the gene variations in the *HSPD1* gene in spastic paraplegia patients has generally been mostly negative. However, in the study by Svenstrup et al., homozygosity for this variation in a sporadic spastic paraplegia patient was the only genetic finding made (Svenstrup et al. 2009), opening the possibility that homozygosity might be disease-associated. As the allele frequency for the p.GLy536Ala variation is rather high in Americans of the European descendents and the Danish population, a rather high number of individuals in these populations are expected to be homozygous for this variation. If it had a significant power of causing spastic paraplegia, this would very likely not have gone unnoticed.

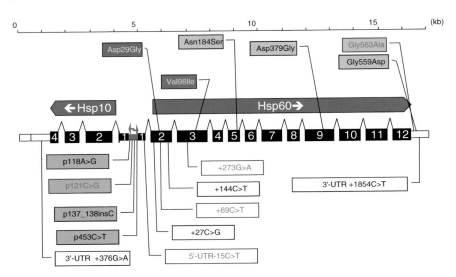

Fig. 12.1 Sequence variations detected in the human *HSPD1 and HSPE1* genes. The position of validated sequence variations in the coding region, 5′ and 3′ UTR's and the promoter are shown. Amino acid replacing variations are shown above the exon gene structure cartoon; the two disease-associated variations are highlighted in *red*. Below the cartoon, silent nucleotide variations and variations in the UTR's of the transcripts are shown. No gene variations were so far detected in the *HSPE1* region

Besides polymorphic SAP's, a couple of further amino acid variations in Hsp60 have been detected by us and by the NHLBI Exome Sequencing Project (http://evs. gs.washington.edu/EVS/). As discussed in detail further below, some of these have been shown to be disease-associated (Magen et al. 2008; Hansen et al. 2002, 2007), whereas for the others it is unclear whether they have any impact as they were coincidentally detected in control individuals or population samples without any records on potential phenotypic effect.

At nucleotide level, there are a couple of silent nucleotide polymorphisms in the *HSPD1* coding region, nucleotide polymorphisms in the 5′UTR, and in the bidirectional promoter (Bross et al. 2007). The promoter polymorphism has been tested in a reporter assay and showed no effect on transcription rates. In the *HSPE1* transcript, another polymorphic site in the 5′UTR has been recorded in the NHLBI Exome Sequencing Project (http://evs.gs.washington.edu/EVS/). Possible consequences of these polymorphic variations for expression, splicing, and mRNA stability remain to be investigated experimentally.

Chapter 13
Type I Chaperonins Are Essential for Cell Viability

Abstract Knock-out of type I chaperonin genes in organisms from bacteria to mice is lethal indicating that these genes are essential and suggesting that variations in these genes possess a high probability to cause deleterious phenotypes.

As discussed above, chaperonins are with the exception of some symbiotic *Mycoplasma* species present in all living cells and knocking out or inactivating of the genes encoding the type I chaperonins in *E. coli* (Fayet et al. 1989), *Saccharomyces cerevisiae* (Cheng et al. 1989), and *Drosophila melanogaster* (Perezgasga et al. 1999) is lethal. Mutations in the gene encoding the zebrafish Hsp60 gene displays a defect in regeneration (Makino et al. 2005); and mutation in the gene encoding a chloroplast type I chaperonin of *Arabidopsis* chloroplast type I chaperonin cause delayed embryonal development (Apuya et al. 2001). All these examples indicate that these genes play very important roles for cellular function and survival suggesting that variations in these genes have a high probability to cause deleterious phenotypes.

Knock-out of mice by introducing a gene trap cassette into intron 2 of the mouse *HSPD1* gene resulted in a lethal phenotype in homozygous knock-out mice (Christensen et al. 2010). Embryos degenerated around day 6.5–7.5 after gestation (Fig. 13.1). Heterozygous knock-out animals were born in the expected ratio, had no apparent phenotype and developed normally although Hsp60 transcript and protein levels were approximately half of those in control animals in liver, heart, brain, and skeletal muscle. This suggests that half levels of the Hsp60 chaperonin are sufficient to maintain all the functions necessary during embryonal and postnatal development. We observed though that crosses between male heterozygotes and wild type females produced a skewed sex ratio with an overrepresentation of males

© The Author(s) 2015
P. Bross, *The Hsp60 Chaperonin*, SpringerBriefs in Molecular Science,
DOI 10.1007/978-3-319-26088-4_13

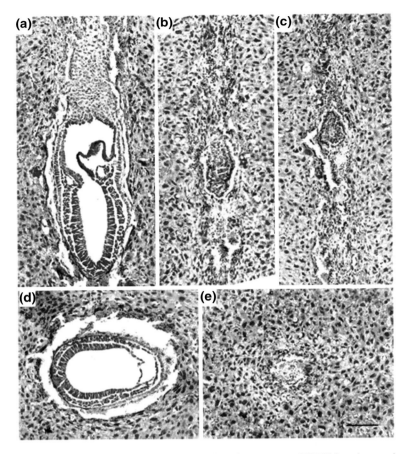

Fig. 13.1 Evidence for degeneration of embryos from homozygous *HSPD1* knock-out mice. HE staining of paraffin sections of five embryos from the same 7.5 dpc litter. **a–c** Longitudinal sections; **d, e** cross section. Each picture was taken at the level where the respective embryo had its largest width. **a, d** Normal-appearing embryos. **b, c**, and **e** Degenerating embryos in different stages of resorption. Scale bar = 100 μm. With kind permission from Springer Science + Business Media: Christensen et al. (2010).

while the sex ratio was at standard level when heterozygous females were crossed with wild type males. This is consistent with the hypothesis that sperms with the Y chromosome move faster than sperms carrying an X chromosome, as the Y chromosome weighs less. Half amounts of Hsp60 in the mitochondria may make this difference in mobility even larger.

Chapter 14
Human Diseases Caused by Genetic Mutations in the Hsp60/Hsp10 System

Abstract Missense mutations in Hsp60 are the cause of very rare inherited neurological diseases. The severity of the disease depends on the degree of functional Hsp60 deficiency.

14.1 Unconfirmed Hsp60 Deficiency-Related Diseases

The first suggestions for human diseases caused by deficiency of the Hsp60/Hsp10 system were published in 1993 (Agsteribbe et al. 1993). At this time, many of the mitochondrial biogenesis mechanisms were only scarcely known. Hsp60 was picked as candidate for the affected function causing this fatal multisystem disorder with deficiency of many metabolic pathways residing in the mitochondrial inner membrane and matrix space because it could have explained the effect on many enzymes. The levels of Hsp60 appeared clearly decreased and the conclusion was that Hsp60 or proteins involved in mitochondrial protein import were carrying mutations. Subsequent genetic analysis by sequencing of the exons, exon–intron borders, and promoter for the *HSPD1* and *HSPE1* genes showed no gene variations that could be responsible for Hsp60 deficiency. In spite of this, the original notion that Hsp60 deficiency was the cause remains uncommented in references from recent publications, e.g., Zhang et al. (2013b), DiMauro and Garone (2011).

14.2 Hereditary Spastic Paraplegia SPG13

The first human disease shown to be associated with mutations in the gene encoding Hsp60 was an autosomal dominantly inherited form of hereditary spastic paraplegia, SPG13. Hereditary spastic paraplegia is a neurological syndrome defined by the clinical finding of bilateral lower extremity weakness and spasticity as the predominant manifestations and a gene mutation as the major causative factor (Fink 2013). The disease can in the first place be diagnosed on basis of a characteristic gait (spastic gait) in connection with lower extremity spasticity and

© The Author(s) 2015
P. Bross, *The Hsp60 Chaperonin*, SpringerBriefs in Molecular Science,
DOI 10.1007/978-3-319-26088-4_14

weakness. The overall prevalence of hereditary spastic paraplegia has in different populations been estimated to be 1–10 per 100,000 individuals. It thus represents the second most prevalent motor neuron disease group (Noreau et al. 2014). Spastic paraplegia primarily affects upper motor neurons that mediate the voluntary movements of the legs. Although the cell bodies of these motor neurons, the so-called Betz cells, are already very big with a diameter of up to one-tenth of a millimeter, the larger part of the cell volume is contributed by the axoplasm due to the extreme length of the axons. The Betz cells are localized in the primary motor cortex and they send their axons through the corticospinal tract to the lower lumbar spinal cord where they enervate secondary motor neurons. The secondary motor neurons also possess long axons that enervate the respective leg muscles.

Post-mortem studies of HSP patients have indicated degeneration of axons in the lateral corticospinal tract with the distal ends being more affected than the proximal ones [reviewed in (Fink 2013)]. Hereditary spastic paraplegia is a clinically and genetically very heterogenous neurological syndrome. The *HSPD1* gene was the sixth gene shown to carry gene variations that were causative for hereditary spastic paraplegia. Knowledge of the full human genome sequence and development of DNA sequencing technology has in the ensuing years promoted discovery and led to the currently identified more than 70 different human genes with causative gene variations (Noreau et al. 2014; Fink 2013, 2014). At the time we found *HSPD1* as disease gene, the human genome structure was only partially known and the discovery was aided by the time-coincidence of our classical mapping of the gene and the description of the SPG13 locus in French kindred with the disease (Fontaine et al. 2000). The SPG13 locus comprises many genes and the trigger for specifically analyzing the *HSPD1* and *HSPE1* genes was the previous finding that SPG7 was caused by the mutations in the mitochondrial inner membrane protease paraplegin. Research in yeast had shown that the yeast homologs of this AAA protease are involved in protein quality control of mitochondrial inner membrane proteins (Langer 2000) and it was therefore suggestive that the mitochondrial chaperonin might be a good candidate as it is involved in protein quality control of mitochondrial proteins. At that time, the conception that mitochondrial protein quality control was a major disease mechanism in hereditary spastic paraplegia was in this instance helpful. It later turned out that mutations in a plethora of genes encoding proteins involved in many different cellular processes can cause hereditary spastic paraplegia.

The special architecture of the upper motor neurons may render them particularly vulnerable and it may be a conceivable explanation for the finding that many different gene defects primarily affect this system. However, a similar case can be made for other neurological disorders and furthermore, different mutations in some of the hereditary spastic paraplegia genes have been shown to cause other diseases like, for the *PLP1* gene that is also a disease gene in Pelizaeus–Merzbacher disease, mutations in the *FAH2* gene that also cause a leukodystrophy and different mutations in the *HSPD1* gene encoding the mitochondrial Hsp60 chaperone cause hereditary spastic paraplegia and a white matter disease. There is thus good reason to analyse the molecular disease mechanisms in detail in order to advance from plausibility considerations to evidence corroborated by experiments.

In a multicellular eukaryotic organism with distinct tissues, mutations leading to deficiency of the mitochondrial type I chaperonin system are limited to mild mutations that retain residual activity of the wild type allele in heterozygous configuration or require considerable residual activity of the mutant chaperonin in a homozygous configuration. More severe or dominant negatively working mutations are not compatible with germ cell survival and embryonic development.

14.2.1 Clinical Phenotype

The clinical phenotypes of hereditary spastic paraplegia are separated into two major subgroups: pure and combined also termed uncomplicated and complicated, respectively. While complex forms are clinically characterized by additional neurological (e.g., ataxia, dementia) and and/or non-neurological manifestations (e.g., facial dysmorphy, macular degeneration) is slowly progressing walking difficulty the major manifestation [discussed in (Finsterer et al. 2012)]. The distinction into pure and complex is not totally consistent as even patients in the same family and carrying the same gene mutation may differ in this category. However, it is a helpful characterization for the first diagnosis. Patients in the large French SPG3 family were characterized as expressing pure spastic paraplegia (Fontaine et al. 2000). In comparison to patients with the most frequent form of autosomal dominant spastic paraplegia SPG4, who carry mutations in the *SPAST* gene, showed similar properties like increased reflexes in the lower limbs and decreased vibration sense; however, increased reflexes in the upper limbs and severe functional handicap were more frequent in the SPG13 family. The average age of onset for SPG13 patients was 39 years.

The c.292G<A mutation in the *HSPD1* gene segregated with the phenotype in most family members; only a few younger mutation carriers did not display symptoms, suggesting that they had not reached the age for disease onset yet. Altogether, a picture of high penetrance and a rather well defined clinical phenotype appears to apply to SPG13.

The c.292G<A variation causes replacement of valine-98 in the Hsp60 precursor protein with isoleucine (p.Val98Ile), i.e., introduces an extra methyl group into the side chain. The valine at this position is highly conserved in type I chaperonins from *Escherichia coli* to humans, however, with the exception of *Caenorhabditis elegans* Hsp60 that actually carries an isoleucine at this position. Valine-98 is situated in the equatorial domain of the protein, buried and forming part of an alpha helix (Fig. 14.1). It is distant from the ATP binding site. Comparing the open and

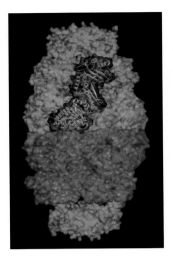

Val98   Asp29

Fig. 14.1 Position of the disease causing Hsp60 mutations in the human Hsp60 structure. *Left* the structure of the human Hsp60 14-mer/Hsp10/14-mer complex based on PDB coordinates 4PJ1 is shown in surface representation. The surface of one Hsp60 subunit is shown in transparent with underlying protein chain as solid ribbon, bound nucleotide as sticks, and the two mutated positions in red space-filling representation. *Right* enlarged view of the highlighted subunit. Representations were produced using Accelrys Discovery Studio Client 4.0

closed conformations seen in the *E. coli* structure of the asymmetric complex, the corresponding valine-74 undergoes some movement when shifting between the open and the closed conformation. Incorporation of the larger isoleucine instead of valine may hamper this movement during the conformational changes.

14.3   MitCHAP60 Disease

In 2008, patients with a fatal white matter disease associated with homozygosity for a p.Asp29Ala missense mutation in the *HSPD1* gene were described. In the large consanguineous kindred the mutation fully segregated with the phenotype (Magen et al. 2008). Heterozygous individuals did not show any symptoms while homozygous patients all suffered of this autosomal-recessive fatal neurodegenerative disorder of early onset. Early disease symptoms included hypotonia, nystagmus, and psychomotor developmental delay, followed by appearance of prominent spasticity, developmental arrest, and regression. Spastic paraplegia was a comorbidity symptom in all patients. Patients usually died within the first two decades, but severity was very variable. MRI/MRS scanning revealed high intensity of the white matter consistent with the complete absence of myelin. Although a very severe disease due to deficiency of a protein highly expressed in all tissues, the symptoms are apparently delimited to nervous tissue.

The mutated asparagine 29 is like the SPG134 mutation localized in the equatorial domain, also distant from the ATP binding site (Fig. 14.1). Replacement of glycine for the charged and more bulky aspartic acid may also disturb inter-subunit interactions or affect conformational movements.

Chapter 15
Molecular Investigations of Disease Mechanisms

Abstract In vitro studies with purified mutant Hsp60 proteins and in vivo studies with gene-modified mice and patient and model cells have revealed some of the molecular mechaisms underlying Hsp60 deficiency in humans.

Given the genetic association of the mutations causing spastic paraplegia SPG13 and MitCHAP60 disease, functional characterization of the mutation effects was performed in a number of ensuing studies. Genetic complementation analysis in the *Escherichia coli* complementation system clearly showed that the Hsp60-p. Val98Ile mutant protein had a functional deficit. *E. coli* bacteria are viable when the endogenous *groEL* and *groES* genes are knocked out and a plasmid driving expression of human Hsp10 and Hsp60 is present (Richardson et al. 2001). However, when the Hsp60 gene carried the p.Val98Ile mutation, cells did not grow (Hansen et al. 2002). This clearly showed that the mutation has a deleterious effect. The same analysis for the MitCHAP60 disease mutation p.Asp29Gly showed that *E. coli* cells lacking the GroEL and GroES genes and expressing human Hsp10 and Hsp60-p.Asp29Gly were viable, however, they were impaired in growth compared to cells expressing wild type Hsp60.

Both the Hsp60-P.Val98Ile and the Hsp60-p.Asp29Gly protein were recombinantly expressed and purified from bacteria. Both mutant proteins formed oligomeric ring structures like the wild type protein (Bross et al. 2008; Parnas et al. 2009). However, the ATPase activities of both proteins were significantly decreased as well as the capacity to refold the model substrate malate dehydrogenase in vitro (shown for the Hsp60-P.Val98Ile mutant protein in Fig. 15.1).

For the p.Asp29Gly mutation which is present together with a wild type allele in SPG13 patients, we investigated whether the mutant Hsp60-p.Val98Ile has a dominant negative effect on wild type Hsp60 when expressed in the same bacterial cell. Assuming that both wild type and mutant Hsp60 are incorporated into Hsp60 ring complexes and given the decreased ATPase and folding activity of the Hsp60-p.Val98Ile protein, one might expect that mixed complexes consisting of wild type and at least one mutant Hsp60 subunit would be less active than complexes consisting of wild type Hsp60 only. This issue was experimentally addressed

P. Bross, *The Hsp60 Chaperonin*, SpringerBriefs in Molecular Science,
DOI 10.1007/978-3-319-26088-4_15

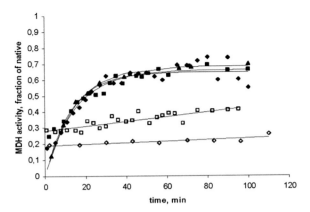

Fig. 15.1 In vitro refolding capacity of Hsp60-p.Val98Ile refolding activity. Refolding of acid-denatured malate dehydrogenase by wild-type Hsp60/Hsp10 (closed diamonds), Hsp60-(p. Gly67Ser)/Hsp10 (a mutant variant that has no effect on activity; *closed boxes*), *E. coli* GroEL/GroES (*closed triangles*), the disease-associated mutation Hsp60-(p.Val98Ile)/Hsp10 (*open boxes*), or bovine serum albumin (*open diamonds*) was measured following regain of malate dehydrogenase enzyme activity over time and expressed as fraction of the activity before denaturation. This research was originally published in *J. Biol. Chem.*: Bross et al. (2008)

by genetically engineered *E. coli* cells to carry two "alleles" encoding the human Hsp10 and Hsp60 proteins: The cells were provided with a plasmid encoding an Hsp10/Hsp60 cDNA operon under control of an IPTG-inducible promoter; the endogenous *groES/groEL* operon was subsequently deleted by homologous recombination as in the complementation experiments described above; thereafter, these cells were transformed with a second separately selectable and replicating plasmid carrying a Hsp10/Hsp60-p.Val98Ile cDNA operon controlled by a arabinose-inducible promoter. In these cells, expression of either of the two human chaperonin operons or both in parallel could be accomplished. As control cells with two wild type cDNA operons and cells with a wild type and a mutant operon with an artificially constructed ATPase-deficient mutant that is expected to have a dominant negative effect when such subunits are incorporated into rings together with wild type subunits were also engineered. Co-induction of two wild type human chaperonin operons or a wild type and a p.Val98Ile mutant containing operon resulted in very similar growth curves. In contrast, induction of both a wild type and an ATPase mutant operon strongly blocked bacterial growth. This suggests that the ATPase deficient Hsp60 has a strong dominant negative effect on chaperonin function whereas the p.Val98Ile mutant Hsp60 does not.

15.1   Cultured Patient Cells

Analysis of SPG13 patient fibroblast and lymphoblastoid cells using LC-MS/MS showed that the Hsp60-p.Val98Ile mutant protein was present in similar amounts as the wild type in these heterozygous cells (Hansen et al. 2008). This suggested that the mutation had no significant effect on folding and stability of the Hsp60-p. Val98Ile protein. Immunostaining of patient fibroblast cells with anti-Hsp60 antibodies showed full co-localization of the Hsp60 signal with the mitochondrial marker mitotracker—red indicating that the mutant Hsp60 protein was correctly targeted to mitochondria (Fig. 15.2). There were no marked differences with respect to mitochondrial volume or gross mitochondrial size and structure in patient cells compared to control fibroblasts. Furthermore, mitochondrial membrane potential in patient lymphoblastoid cells and control lymphoblastoid cells was similar. Viability measured using the MTT assay under normal and oxidative stress conditions (H_2O_2 treatment) showed that patient cells were indistinguishable from respective controls in this analysis.

The Hsp60/Hsp10 chaperonin, together with other matrix chaperones like the mitochondrial Hsp70 system and certain matrix proteases, involved in mitochondrial protein quality control. Protein quality control is tuned to the demands by adjusting the levels of the different components. We thus asked if there were compensatory responses based on the regulatory circuits that adapt the cell to

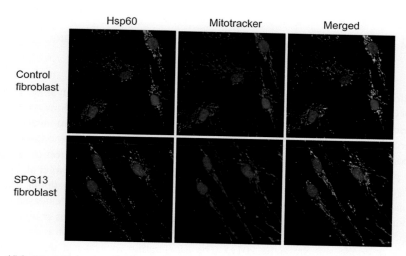

Fig. 15.2 Intracellular localization of Hsp60 protein in SPG13 patient fibroblast cells and controls. Heterozygous mutant fibroblasts from a patient and wild type control cells mounted on slides were fixed and Hsp60 protein was visualized using Hsp60 primary antibody and conjugated secondary antibody (*green*), cell nuclei were visualized with Hoechst (*blue*) and mitochondria with mitotracker (*red*). Mass spectrometry had shown that the mutant Hsp60-p.Val98Ile protein was present in similar amounts as the wild type protein in patient fibroblasts. Reprinted from: Hansen et al. (2008), with permission from Elsevier

decreased Hsp60 chaperone function. Analysis of transcript and protein levels revealed that SPG13 patient lymphoblastoid cells displayed lower expression of the matrix proteases Lon and ClpP both at transcript and protein levels. The subunits for the inner membrane AAA-protease, the Hsp60, mtHsp70 chaperones, and their co-chaperones displayed similar levels in patient and control lymphoblastoid cells. This result may suggest that patient cells adapt by decreasing the capacity of the degradative machinery of the protein quality control in the matrix to reestablish a balance that gives proteins undergoing folding a better chance to fold by reducing the probability to be attacked by these proteases while undergoing folding. How Hsp60/10 functional shortage would down-regulate transcription of the *LON* and *CLPP* genes is unclear. It is commonly assumed that the mitochondrial protein quality control genes are regulated via certain transcription factors that bind to CHOP, MURE, and HSF binding sites in the respective promoters (see above), but it is unclear how transcription could be fine-tuned so that balances between the different components can be adapted. The all-up or all-down concept for high and low stress conditions, respectively, may have to be reevaluated. However, the experimental findings are based on cells from a single patient. Proteome analyses of brain, spinal cord, and liver mitochondria in the heterozygous *Hspd1* mouse model for SPG13 (see below) did not show significant alterations in the levels of the Lon and ClpP protease levels. The observed decreased levels of the LON and CLPP proteases must therefore be taken with caution and additional studies have to be performed.

15.2 Mouse Model for Hereditary Spastic Paraplegia SPG13

As described above, homozygous knock-out of the *Hspd1* gene in mice caused embryonal lethality while heterozygous knock out animals did not show an evident severe early onset phenotype. However, more detailed long-term analyses of heterozygous *Hspd1* knock-out mice indicated that they represent a good model for hereditary spastic paraplegia SPG13 showing late onset motor neuron defect and thus phenocopying some of the features observed in humans (Magnoni et al. 2013). Measurements of motor functions in these mice showed that heterozygous mutant animals from age 1 year on displayed decreased clasping reflex and decreased score for hind limb extension during tail suspension compared to wild type control mice (Fig. 15.3). This is indicative of a disconnection between premotor and motor cortex (Lalonde and Strazielle 2011). The accelerating rotarod test that measures coordination, balance, and motor learning abilities likewise showed significant deficits from age 1 year onwards in heterozygous mutant versus control mice [Fig. 15.3 (Magnoni et al. 2013)].

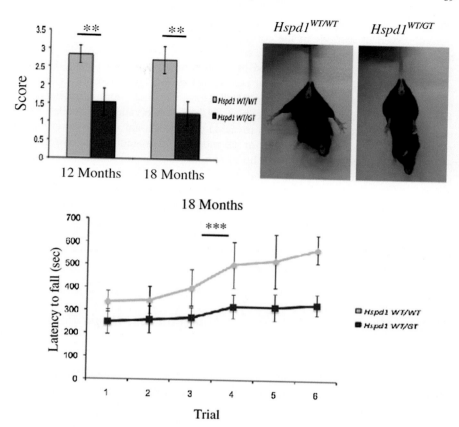

Fig. 15.3 Motor phenotype characterization of heterozygous *Hspd1* knock-out mice (*Hspd1WT/GT*). *Upper panel* Quantitation of the hind-limb extension reflex at 12 and 18 months of age with statistical significance compared to wild type littermates. On the *left,* examples of a *Hspd1WT/WT* mouse showing normal extension reflex and a *Hspd1WT/GT* mouse displaying extension reflex impairment after tail suspension. *Lower panel* Quantitation of Rotarod performance of *Hspd1WT/GT* and wild-type mice at 18 months of age. Data are presented as means ± SD, ANOVA, (***p b 0.001), (*n* = 24). Reprinted from: Magnoni et al. (2013), with permission from Elsevier

15.2.1   *Hspd1(+/−) Mice Display Morphological Changes in the Corticospinal Tract*

Histological analyses of young (3 months) and old (18 months) mice indicated that the spinal cord was a major site of morphological differences between mutant and control mice. Morphological evaluation of the light microscopic preparations of gastrocnemius muscle indicated a decrease of fiber size only at 18 months of age. Muscle fibers with centrally placed nuclei were sporadically seen at 3 months of age. No evident anomalies were observed in liver. Clear morphological changes were observed in sections of the corticospinal tracts in the spinal cord of mutant mice indicating an increased presence of large caliber axons at the age of 3 months. This difference was further enhanced in 18 months old mutant animals. Electron microscopy analyses of corticospinal tract axons showed the presence of swollen mitochondria already in 3 months old mutant mice and more pronounced in 18 months old mutant mice. In agreement with the observation of swollen mitochondria, morphometric quantitation revealed that the axonal volume taken up by mitochondria was clearly increased in 18 months old mutant mice. Histological analysis of the anterior and posterior horns at both cervical and lumbar levels of the spinal cord of mutant mice at 3 and 18 months showed no depletion or other cytological changes of anterior horn cells, suggesting that Hsp60 haploinsufficiency does not impact on a broad selection of neurological tissues, but is rather confined to the axons of the corticospinal tract. Taken together, these data support the notion that heterozygous *Hspd1* knock-out mice represent a valuable model for spastic paraplegia SPG13 that can be used to study the molecular mechanisms underlying the development of the disease phenotype.

15.2.2   *Hspd1(+/−) Mice Display Mitochondrial Respiratory Chain Deficiency*

Studies of mitochondrial proteomes indicated that heterozygous mutant mice showed differences in the expression of respiratory chain subunit proteins. These were most pronounced in brain cortex mitochondria, the region where the cell bodies are localized and which sends their very long axons through the corticospinal tract to the secondary motor neurons. Measurements of ATP production in mitochondria isolated from brain cortex and spinal cord showed that respiratory chain activity was decreased in both types of mitochondria from mutant animals at ages 3 and 18 months. The decreased activity could be limited to complex III and to one of the subunits of complex III, the UQCRC1 protein was shown to be consistently decreased in mutant mice by MS analyses of mitochondrial proteomes. Complex III decrease was confirmed by Blue Native PAGE of respiratory chain complexes and activity staining. Western blotting was also consistent with a decrease of the UQCRC1 protein subunit of complex III. As *Uqcrc1* transcript levels at the same time were increased, the decreased UQCRC1 protein levels in mouse mitochondria with half levels of Hsp60 are consistent with the notion that— due to limiting chaperonin help—the UQCRC1 protein does not fold efficiently and is consequently degraded by protein quality control proteases in the mitochondrial matrix. The UQCRC1 protein is the core 1 subunit that during assembly of complex III first interacts with the core 2 subunit (Uqcrc2) and is then in the inner mitochondrial membrane assembled with the catalytic subunits to a monomeric complex after which two complexes dimerize and incorporate the Rieske iron sulphur and Qcr10 proteins (Ghezzi and Zeviani 2012).

15.2.3   Hspd1(+/−) Mice Display Increased Protein Carbonylation and That Is at Least in Part Due to Defective Folding of the MN-Dependent Superoxide Dismutase SOD2

Oxidative stress, the increased production of reactive oxygen species inside mitochondria is a hallmark of most disorders in which mitochondrial dysfunction plays a role. A major source for generation of reactive oxygen species is defective electron transfer of complex I and complex III in the respiratory chain resulting in transfer of leaking electrons to molecular oxygen and thus generating superoxide. Superoxide is converted to the less reactive hydrogen peroxide by superoxide dismutases and hydrogen peroxide is further detoxified by an array of enzymes with glutathione peroxidase and peroxyredoxins playing major roles inside mitochondria. Given the finding that complex III activity and levels were decreased in $Hspd1(\pm)$ mice, one would expect an increased superoxide production by electrons leaking from complex I due to partial blockage of their further transfer to complex III. A result of this should be a higher level of reactive oxygen species and increased damage to macromolecules. Assessment of protein carbonylation levels using oxyblot and an ELISA-based method in mitochondrial preparations from brain cortex and spinal cord of heterozygous mutant and control mice indeed showed an increase in carbonylated proteins (Magnoni et al. 2014). Proteomic analysis indicated decreased levels of the matrix superoxide dismutase SOD2 that detoxifies the very reactive superoxide to hydrogen peroxide. This finding was validated by analysing SOD2 transcript levels that turned out to be unchanged in these tissues of mutant mice compared to controls. Direct measurement of SOD2 enzyme activity and Western blotting confirmed that SOD2 activity and protein levels were decreased in spinal cord and brain cortex mitochondria of mutant mice. Taken together this suggested that decrease in SOD2 amounts and activity was due to a posttranscriptional mechanism. Given that mutant mice display reduced levels of the Hsp60 chaperonin, it was attractive to speculate that SOD2 folding and assembly is dependent on the Hsp60/Hsp10 chaperone system and that limiting amounts of Hsp60/Hsp10 complexes will cause SOD2 polypeptide chains being trapped in folding intermediates that are degraded by protein quality control proteases in the mitochondrial matrix.

To further explore this, it was investigated whether SOD2 protein could be co-immunoprecipitated using antibodies directed against Hsp60 complexes. This was both performed after in vitro translation of the SOD2 protein followed by import into mitochondria isolated from brain of heterozygous mutant and control mice and by immunoprecipitation of extracts from mitochondria isolated from brain cortex and spinal cord. In both types of experiment SOD2 protein was brought down by the Hsp60 antibodies suggesting that the two proteins in fact physically interact.

Native PAGE analysis of in vitro synthesized SOD2 after importing it into mitochondria from brain of mutant and wild type mice showed dramatically decreased levels of SOD2 dimers in the mutant mitochondria. Consistent with this, the SOD2 band co-migrating with Hsp60 complexes was increased in mitochondria from mutant mice. This suggests that folding of SOD2 is inefficient upon Hsp60 deficiency and that this results in degradation of SOD2 folding intermediates and a low superoxide dismutase activity in the mitochondrial matrix. This may be the foundation for increased damage to macromolecules mediated by higher superoxide levels which were observed as increased protein carbonylation (see above).

Altogether, this suggests a coupling of the control of superoxide production to the capacity and load of the Hsp60/Hsp10 chaperonin. Under stress conditions, when the Hsp60/Hsp10 system is struggling with a high load of unfolded and misfolded proteins in the mitochondrial matrix, folding of the very highly chaperonin-dependent SOD2 might be impaired resulting in decreased SOD2 activity and increased superoxide levels. The increased superoxide levels may then start a signaling cascade that reestablishes proteostasis in the matrix by upregulating and activating protein quality control components. In yeast, Cabiscol et al. described a coupling of Hsp60 levels to the levels of oxidative stress which is consistent with this notion (Cabiscol et al. 2002).

Chapter 16
Outlook

Abstract The future research in chaperonin deficiency diseases can contribute important knowledge for the many diseases in which mitochondrial dysfunction plays an important role.

Research on the mammalian Hsp60/Hsp10 complex has come a long way and we today understand many of its basic mechanisms and biological roles. However, as always in science, there are still many open questions. For the genetic deficiencies in humans caused by mutations in the gene encoding Hsp60, it is still puzzling why such mutations affect primarily neuronal tissues and have no phenotypic impact on other tissues like heart and skeletal muscle, which also are very much dependent on mitochondria. The connection between Hsp60 deficiency and defective myelination remains also to be elucidated.

As far as Hsp60 is concerned, we know that Hsp60 is essential for viability of all mitochondria eukaryotic cells. It is involved in the folding of a long series of mitochondrial matrix proteins including respiratory chain subunits, respiratory chain assembly factors, fatty acid oxidation, and other metabolic as well as antioxidant enzymes. Hsp60 deficiency affects folding of many nuclear encoded matrix and inner membrane proteins but also some mitochondrial DNA encoded respiratory chain subunits.

Future research will have to define the Hsp60/Hsp10 client proteins and identify those mitochondrial proteins that are particularly dependent on this chaperonin system. Like apparently SOD2, folding of these proteins will be impaired first upon Hsp60/Hsp10 deficiency caused by stress conditions or due to genetic deficiencies. An important, and so far not very much addressed, aspect is to assess those proteins which not only require Hsp60/Hsp10 assistance during their folding after importing into mitochondria, but also those that require repeated refolding assistance [the so-called conformational maintenance (Houry et al. 1999), due to conformational lability. Such proteins may be sensors for the cell to determine the proteostasis capacity and play an important role for signaling of stress response mechanisms. Finally, as the connection between insulin signaling, Hsp60 expression regulation and mitochondrial function maintenance indicate, the Hsp60/Hsp10 system may be

© The Author(s) 2015
P. Bross, *The Hsp60 Chaperonin*, SpringerBriefs in Molecular Science,
DOI 10.1007/978-3-319-26088-4_16

a key hub/regulator of mitochondrial function that is addressed by intracellular signaling circuits, as well as by endocrine signaling. A better understanding of these mechanisms and options to modulate it could have great potential for treatment of the ever-growing number of diseases where mitochondrial dysfunction is an important part of the pathogenetic mechanism (Nunnari and Suomalainen 2012; Viscomi et al. 2015).

Bibliography

Adachi N, Lieber MR (2002) Bidirectional gene organization: a common architectural feature of the human genome. Cell 109(7):807–809

Agsteribbe E, Huckriede A, Veenhuis M, Ruiters MH, Niezen-Koning KE, Skjeldal OH, Skullerud K, Gupta RS, Hallberg R, van Diggelen OP et al (1993) A fatal, systemic mitochondrial disease with decreased mitochondrial enzyme activities, abnormal ultrastructure of the mitochondria and deficiency of heat shock protein 60. Biochem Biophys Res Commun 193(1):146–154

Aldridge JE, Horibe T, Hoogenraad NJ (2007) Discovery of genes activated by the mitochondrial unfolded protein response (mtUPR) and cognate promoter elements. PLoSONE 2(9):e874

Andresen BS, Bross P, Udvari S, Kirk J, Gray G, Kmoch S, Chamoles N, Knudsen I, Winter V, Wilcken B, Yokota I, Hart K, Packman S, Harpey JP, Saudubray JM, Hale DE, Bolund L, Kølvraa S, Gregersen N (1997) The molecular basis of medium-chain acyl-CoA dehydrogenase (MCAD) deficiency in compound heterozygous patients: is there correlation between genotype and phenotype? Hum Mol Genet 6(5):695–707

Andresen BS, Dobrowolski SF, O'Reilly L, Muenzer J, McCandless SE, Frazier DM, Udvari S, Bross P, Knudsen I, Banas R, Chace DH, Engel P, Naylor EW, Gregersen N (2001) Medium-chain acyl-CoA dehydrogenase (MCAD) mutations identified by MS/MS-based prospective screening of newborns differ from those observed in patients with clinical symptoms: identification and characterization of a new, prevalent mutation that results in mild MCAD deficiency. Am J Hum Genet 68(6):1408–1418

Anfinsen CB (1973) Principles that govern the folding of protein chains. Science 181:223–230

Angelucci F, Saccoccia F, Ardini M, Boumis G, Brunori M, Di LL, Ippoliti R, Miele AE, Natoli G, Scotti S, Bellelli A (2013) Switching between the alternative structures and functions of a 2-Cys peroxiredoxin, by site-directed mutagenesis. J Mol Biol 425(22):4556–4568

Apuya NR, Yadegari R, Fischer RL, Harada JJ, Zimmerman JL, Goldberg RB (2001) The Arabidopsis embryo mutant schlepperless has a defect in the chaperonin-60alpha gene. Plant Physiol 126(2):717–730

Azia A, Unger R, Horovitz A (2012) What distinguishes GroEL substrates from other Escherichia coli proteins? Febs J 279(4):543–550. doi:10.1111/j.1742-4658.2011.08458.x

Balch WE, Morimoto RI, Dillin A, Kelly JW (2008) Adapting proteostasis for disease intervention. Science 319(5865):916–919

Baneyx F (1999) Recombinant protein expression in Escherichia coli. Curr Opin Biotechnol 10 (5):411–421

Bogumil D, Dagan T (2012) Cumulative impact of chaperone-mediated folding on genome evolution. Biochemistry-Us 51(50):9941–9953

Bonior J, Jaworek J, Konturek SJ, Pawlik WW (2006) Leptin is the modulator of HSP60 gene expression in AR42J cells. J Physiol Pharmacol 57(Suppl 7):135–143

Boshoff A (2015) Chaperonin-Co-chaperonin Interactions. Subcell Biochem 78:153–178.

© The Author(s) 2015
P. Bross, *The Hsp60 Chaperonin*, SpringerBriefs in Molecular Science,
DOI 10.1007/978-3-319-26088-4

Braig K, Otwinowski Z, Hegde R, Boisvert DC, Joachimiak A, Horwich AL, Sigler PB (1994) The crystal structure of the bacterial chaperonin GroEL at 2.8 angstrom. Nature 371:578–586

Brocchieri L, Karlin S (2000) Conservation among HSP60 sequences in relation to structure, function, and evolution. Protein Sci 9(3):476–486

Bross P, Andresen BS, Winter V, Krautle F, Jensen TG, Nandy A, Kolvraa S, Ghisla S, Bolund L, Gregersen N (1993) Co-overexpression of bacterial GroESL chaperonins partly overcomes non-productive folding and tetramer assembly of *E. coli*-expressed human medium-chain acyl-CoA dehydrogenase (MCAD) carrying the prevalent disease-causing K304E mutation. Biochim Biophys Acta 1182(3):264–274

Bross P, Jespersen C, Jensen TG, Andresen BS, Kristensen MJ, Winter V, Nandy A, Kräutle F, Ghisla S, Bolund L, Kim JJP, Gregersen N (1995) Effects of two mutations detected in medium chain acyl-CoA dehydrogenase (MCAD)-deficient patients on folding, oligomer assembly, and stability of MCAD enzyme. J Biol Chem 270:10284–10290

Bross P, Li Z, Hansen J, Hansen JJ, Nielsen MN, Corydon TJ, Georgopoulos C, Ang D, Lundemose JB, Niezen-Koning K, Eiberg H, Yang H, Kolvraa S, Bolund L, Gregersen N (2007) Single-nucleotide variations in the genes encoding the mitochondrial Hsp60/Hsp10 chaperone system and their disease-causing potential. J Hum Genet 52(1):56–65

Bross P, Naundrup S, Hansen J, Nielsen MN, Christensen JH, Kruhoffer M, Palmfeldt J, Corydon TJ, Gregersen N, Ang D, Georgopoulos C, Nielsen KL (2008) The HSP60-(P.val98ile) mutation associated with hereditary spastic paraplegia SPG13 compromises chaperonin function both in vitro and in vivo. J Biol Chem 283(23):15694–15700

Cabiscol E, Belli G, Tamarit J, Echave P, Herrero E, Ros J (2002) Mitochondrial Hsp60, resistance to oxidative stress and the labile iron pool are closely connected in *Saccharomyces cerevisiae*. JBiolChem 277(46):44531–44538

Calderwood SK, Khaleque MA, Sawyer DB, Ciocca DR (2006) Heat shock proteins in cancer: chaperones of tumorigenesis. Trends Biochem Sci 31:194–172

Cappello F, Conway de Macario E, Marasa L, Zummo G, Macario AJ (2008) Hsp60 expression, new locations, functions and perspectives for cancer diagnosis and therapy. Cancer Biol Ther 7 (6):801–809

Chapman E, Farr GW, Usaite R, Furtak K, Fenton WA, Chaudhuri TK, Hondorp ER, Matthews RG, Wolf SG, Yates JR, Pypaert M, Horwich AL (2006) Global aggregation of newly translated proteins in an *Escherichia coli* strain deficient of the chaperonin GroEL. Proc Natl Acad Sci USA 103(43):15800–15805

Chaudhuri TK, Farr GW, Fenton WA, Rospert S, Horwich AL (2001) GroEL/GroES-mediated folding of a protein too large to be encapsulated. Cell 107(2):235–246

Chen DH, Song JL, Chuang DT, Chiu W, Ludtke SJ (2006) An expanded conformation of single-ring GroEL-GroES complex encapsulates an 86 kDa substrate. Structure 14(11):1711–1722

Chen XY, Sullivan DS, Huffaker TC (1994) Two yeast genes with similarity to TCP-1 are required for microtubule and actin function in vivo. Proc Natl Acad Sci USA 91:9111–9115

Cheng MY, Hartl FU, Horwich AL (1990) The mitochondrial chaperonin hsp60 is required for its own assembly. Nature 348(6300):455–458

Cheng MY, Hartl FU, Martin J, Pollock RA, Kalousek F, Neupert W, Hallberg EM, Hallberg RL, Horwich AL (1989) Mitochondrial heat-shock protein hsp60 is essential for assembly of proteins imported into yeast mitochondria. Nature 337(6208):620–625

Christensen JH, Nielsen MN, Hansen J, Fuchtbauer A, Fuchtbauer EM, West M, Corydon TJ, Gregersen N, Bross P (2010) Inactivation of the hereditary spastic paraplegia-associated Hspd1 gene encoding the Hsp60 chaperone results in early embryonic lethality in mice. Cell Stress Chaperones 15(6):851–863

Clare DK, Bakkes PJ, van Heerikhuizen H, van der Vies SM, Saibil HR (2009) Chaperonin complex with a newly folded protein encapsulated in the folding chamber. Nature 457 (7225):107–110

Clare DK, Vasishtan D, Stagg S, Quispe J, Farr GW, Topf M, Horwich AL, Saibil HR (2012) ATP-triggered conformational changes delineate substrate-binding and -folding mechanics of the GroEL chaperonin. Cell 149(1):113–123

Clark PL (2004) Protein folding in the cell: reshaping the folding funnel. Trends Biochem Sci 29 (10):527–534

Cloutier P, Coulombe B (2013) Regulation of molecular chaperones through post-translational modifications: decrypting the chaperone code. Biochim Biophys Acta 1829(5):443–454

Cloutier P, Lavallee-Adam M, Faubert D, Blanchette M, Coulombe B (2013) A newly uncovered group of distantly related lysine methyltransferases preferentially interact with molecular chaperones to regulate their activity. PLoS Genet 9(1):e1003210

Cong Y, Baker ML, Jakana J, Woolford D, Miller EJ, Reissmann S, Kumar RN, Redding-Johanson AM, Batth TS, Mukhopadhyay A, Ludtke SJ, Frydman J, Chiu W (2010) 4.0-A resolution cryo-EM structure of the mammalian chaperonin TRiC/CCT reveals its unique subunit arrangement. Proc Natl Acad Sci USA 107(11):4967–4972

Corsepius NC, Lorimer GH (2013) Measuring how much work the chaperone GroEL can do. Proc Natl Acad Sci USA 110(27):E2451–E2459

Corydon TJ, Hansen J, Bross P, Jensen TG (2005) Down-regulation of Hsp60 expression by RNAi impairs folding of medium-chain acyl-CoA dehydogenase wild-type and disease-associated proteins. Mol Genet Metab 85(4):260–270

Csermely P (1999) Chaperone-percolator model: a possible molecular mechanism of anfinsen-cage-type chaperones. BioEssays 21(11):959–965

Cunningham F, Amode MR, Barrell D, Beal K, Billis K, Brent S, Carvalho-Silva D, Clapham P, Coates G, Fitzgerald S, Gil L, Giron CG, Gordon L, Hourlier T, Hunt SE, Janacek SH, Johnson N, Juettemann T, Kahari AK, Keenan S, Martin FJ, Maurel T, McLaren W, Murphy DN, Nag R, Overduin B, Parker A, Patricio M, Perry E, Pignatelli M, Riat HS, Sheppard D, Taylor K, Thormann A, Vullo A, Wilder SP, Zadissa A, Aken BL, Birney E, Harrow J, Kinsella R, Muffato M, Ruffier M, Searle SM, Spudich G, Trevanion SJ, Yates A, Zerbino DR, Flicek P (2015) Ensembl 2015. Nucleic Acids Res 43 (Database issue): D662–669

Dekker C, Willison KR, Taylor WR (2011) On the evolutionary origin of the chaperonins. Proteins 79(4):1172–1192

Dill KA, Chan HS (1997) From levinthal to pathways to funnels. Nat Struct Biol 4(1):10–19

DiMauro S, Garone C (2011) Metabolic disorders of fetal life: glycogenoses and mitochondrial defects of the mitochondrial respiratory chain. SeminFetal Neonatal Med 16(4):181–189

Ditzel L, Lowe J, Stock D, Stetter KO, Huber H, Huber R, Steinbacher S (1998) Crystal structure of the thermosome, the archaeal chaperonin and homolog of CCT. Cell 93(1):125–138

Dubaquie Y, Looser R, Fünfschilling U, Jenö P, Rospert S (1998) Identification of in vivo substrates of the yeast mitochondrial chaperonins reveals overlapping but non-identical requirement for hsp60 and hsp10. EMBO J 17(20):5868–5876

Ellis RJ, Hartl FU (1996) Protein folding in the cell: competing models of chaperonin function. FASEB J 10(1):20–26

Ellis RJ, Hemmingsen SM (1989) Molecular chaperones—proteins essential for the biogenesis of some macromolecular structures. Trends Biochem Sci 14(8):339–342

Ewalt KL, Hendrick JP, Houry WA, Hartl FU (1997) In vivo observation of polypeptide flux through the bacterial chaperonin system. Cell 90(3):491–500

Fares MA, Moya A, Barrio E (2005) Adaptive evolution in GroEL from distantly related endosymbiotic bacteria of insects. J Evol Biol 18(3):651–660

Fares MA, Ruiz-Gonzalez MX, Moya A, Elena SF, Barrio E (2002) Endosymbiotic bacteria: GroEL buffers against deleterious mutations. Nature 417(6887):398

Farr GW, Fenton WA, Chaudhuri TK, Clare DK, Saibil HR, Horwich AL (2003) Folding with and without encapsulation by cis- and trans-only GroEL-GroES complexes. EMBO J 22(13): 3220–3230

Fayet O, Ziegelhoffer T, Georgopoulos C (1989) The groES and groEL heat shock gene products of *Escherichia coli* are essential for bacterial growth at all temperatures. J Bacteriol 171 (3):1379–1385

Fei X, Ye X, LaRonde NA, Lorimer GH (2014) Formation and structures of GroEL: GroES2 chaperonin footballs, the protein-folding functional form. Proc Natl Acad Sci USA 111 (35):12775–12780

Fink JK (2013) Hereditary spastic paraplegia: clinico-pathologic features and emerging molecular mechanisms. Acta Neuropathol 126(3):307–328

Fink JK (2014) Hereditary spastic paraplegia: clinical principles and genetic advances. Semin Neurol 34(3):293–305

Finsterer J, Loscher W, Quasthoff S, Wanschitz J, Auer-Grumbach M, Stevanin G (2012) Hereditary spastic paraplegias with autosomal dominant, recessive, X-linked, or maternal trait of inheritance. J Neurol Sci 318(1–2):1–18

Fontaine B, Davoine CS, Dürr A, Paternotte C, Feki I, Weissenbach J, Hazan J, Brice A (2000) A new locus for autosomal dominant pure spastic paraplegia, on chromosome 2q24-q34. Am J Hum Genet 66(2):702–707

Fujiwara K, Ishihama Y, Nakahigashi K, Soga T, Taguchi H (2010) A systematic survey of in vivo obligate chaperonin-dependent substrates. EMBO J 29(9):1552–1564

Georgescauld F, Popova K, Gupta AJ, Bracher A, Engen JR, Hayer-Hartl M, Hartl FU (2014) GroEL/ES chaperonin modulates the mechanism and accelerates the rate of TIM-barrel domain folding. Cell 157(4):922–934

Georgopoulos CP, Hendrix RW, Kaiser AD, Wood WB (1972) Role of the host cell in bacteriophage morphogenesis: effects of a bacterial mutation on T4 head assembly. Nat New Biol 239(89):38–41

Ghezzi D, Zeviani M (2012) Assembly factors of human mitochondrial respiratory chain complexes: physiology and pathophysiology. Adv Exp Med Biol 748:65–106

Goloubinoff P, Christeller JT, Gatenby AA, Lorimer GH (1989) Reconstitution of active dimeric ribulose bisphosphate carboxylase from an unfoleded state depends on two chaperonin proteins and Mg-ATP. Nature 342(6252):884–889

Goloubinoff P, Gatenby AA, Lorimer GH (1989) GroE heat-shock proteins promote assembly of foreign prokaryotic ribulose bisphosphate carboxylase oligomers in *Escherichia coli*. Nature 337(6202):44–47

Goltermann L, Good L, Bentin T (2013) Chaperonins fight aminoglycoside-induced protein misfolding and promote short-term tolerance in *Escherichia coli*. J Biol Chem 288(15): 10483–10489

Gordon CL, Sather SK, Casjens S, King J (1994) Selective in vivo rescue by GroEL/ES of thermolabile folding intermediates to phage P22 structural proteins. J Biol Chem 269 (45):27941–27951

Gottesman S, Wickner S, Maurizi M (1997) Protein quality control: triage by chaperones and proteases. Genes Dev 11(7):815–823

Grzechnik P, Tan-Wong SM, Proudfoot NJ (2014) Terminate and make a loop: regulation of transcriptional directionality. Trends Biochem Sci 39(7):319–327

Guisbert E, Yura T, Rhodius VA, Gross CA (2008) Convergence of molecular, modeling, and systems approaches for an understanding of the *Escherichia coli* heat shock response. Microbiol Mol Biol Rev 72(3):545–554

Gupta AJ, Haldar S, Milicic G, Hartl FU, Hayer-Hartl M (2014) Active cage mechanism of chaperonin-assisted protein folding demonstrated at single-molecule level. J Mol Biol 426 (15):2739–2754)

Gupta S, Knowlton AA (2002) Cytosolic heat shock protein 60, hypoxia, and apoptosis. Circulation 106(21):2727–2733

Hansen J, Corydon TJ, Palmfeldt J, Durr A, Fontaine B, Nielsen MN, Christensen JH, Gregersen N, Bross P (2008) Decreased expression of the mitochondrial matrix proteases Lon and ClpP in cells from a patient with hereditary spastic paraplegia (SPG13). Neuroscience 153 (2):474–482

Hansen J, Gregersen N, Bross P (2005) Differential handling of wild-type medium-chain acyl-CoA dehydrogenase (MCAD) and a disease-causing variant by E. coli protein quality control proteases. Biochem Biophys Res Commun 333:1160–1170

Hansen J, Svenstrup K, Ang D, Nielsen MN, Christensen JH, Gregersen N, Nielsen JE, Georgopoulos C, Bross P (2007) A novel mutation in the HSPD1 gene in a patient with hereditary spastic paraplegia. J Neurol 254(7):897–900

Hansen JJ, Bross P, Westergaard M, Nielsen MN, Eiberg H, Børglum AD, Mogensen J, Kristiansen K, Bolund L, Gregersen N (2003) Genomic structure of the human mitochondrial chaperonin genes: HSP60 and HSP10 are localised head to head on chromosome 2 separated by a bidirectional promoter. Hum Genet 112(1):71–77

Hansen JJ, Dürr A, Cournu-Rebeix I, Georgopoulos C, Ang D, Nielsen MN, Davoine CS, Brice A, Fontaine B, Gregersen N, Bross P (2002) Hereditary spastic paraplegia SPG13 is associated with a mutation in the gene encoding the mitochondrial chaperonin Hsp60. Am J Hum Genet 70(5):1328–1332

Haynes CM, Ron D (2010) The mitochondrial UPR—protecting organelle protein homeostasis. J Cell Sci 123(Pt 22):3849–3855

Hemmingsen SM (1992) What is a chaperonin? [letter]. Nature 357(6380):650

Hemmingsen SM, Woolford C, van der Vies SM, Tilly K, Dennis DT, Georgopoulos CP, Hendrix RW, Ellis RJ (1988) Homologous plant and bacterial proteins chaperone oligomeric protein assembly. Nature 333(6171):330–334

Henderson B, Fares M, Lund PA (2013) Chaperonin 60: a paradoxical, evolutionarily conserved protein family with multiple moonlighting functions. Biol Rev Camb Philos Soc 88(4): 955–987

Hendrix RW (1979) Purification and properties of groE, a host protein involved in bacteriophage assembly. J Mol Biol 129(3):375–392

Hewamadduma CA, Kirby J, Kershaw C, Martindale J, Dalton A, McDermott CJ, Shaw PJ (2008) HSP60 is a rare cause of hereditary spastic paraparesis, but may act as a genetic modifier. Neurology 70(19):1717–1718

Hirtreiter AM, Calloni G, Forner F, Scheibe B, Puype M, Vandekerckhove J, Mann M, Hartl FU, Hayer-Hartl M (2009) Differential substrate specificity of group I and group II chaperonins in the archaeon Methanosarcina mazei. Mol Microbiol 74(5):1152–1168

Hohfeld J, Hartl FU (1994) Role of the chaperonin cofactor Hsp10 in protein folding and sorting in yeast mitochondria. J Cell Biol 126:305–315

Houry WA, Frishman D, Eckerskorn C, Lottspeich F, Hartl FU (1999) Identification of in vivo substrates of the chaperonin GroEL. Nature 402(6758):147–154

Hwang YJ, Lee SP, Kim SY, Choi YH, Kim MJ, Lee CH, Lee JY, Kim DY (2009) Expression of heat shock protein 60 kDa is upregulated in cervical cancer. Yonsei Med J 50(3):399–406

Illergard K, Ardell DH, Elofsson A (2009) Structure is three to ten times more conserved than sequence—a study of structural response in protein cores. Proteins 77(3):499–508

Ishimoto T, Fujiwara K, Niwa T, Taguchi H (2014) Conversion of a chaperonin GroEL-independent protein into an obligate substrate. J Biol Chem 289(46):32073–32080

Itoh H, Kobayashi R, Wakui H, Komatsuda A, Ohtani H, Miura AB, Otaka M, Masamune O, Andoh H, Koyama K, Sato Y, Tashima Y (1995) Mammalian 60-kDa stress protein (chaperonin homolog)—identification, biochemical properties, and localization. J Biol Chem 270:13429–13435

Itoh H, Komatsuda A, Ohtani H, Wakui H, Imai H, Sawada KI, Otaka M, Ogura M, Suzuki A, Hamada F (2002) Mammalian HSP60 is quickly sorted into the mitochondria under conditions of dehydration. Eur J Biochem 269(23):5931–5938

Kalisman N, Adams CM, Levitt M (2012) Subunit order of eukaryotic TRiC/CCT chaperonin by cross-linking, mass spectrometry, and combinatorial homology modeling. Proc Natl Acad Sci USA 109(8):2884–2889

Kalisman N, Schroder GF, Levitt M (2013) The crystal structures of the eukaryotic chaperonin CCT reveal its functional partitioning. Structure 21(4):540–549

Karlin S, Brocchieri L (2000) Heat shock protein 60 sequence comparisons: duplications, lateral transfer, and mitochondrial evolution. Proc Natl Acad Sci USA 97(21):11348–11353

Keppel F, Rychner M, Georgopoulos C (2002) Bacteriophage-encoded cochaperonins can substitute for *Escherichia coli*'s essential GroES protein. Embo Reports 3(9):893–898

Kerner MJ, Naylor DJ, Ishihama Y, Maier T, Chang HC, Stines AP, Georgopoulos C, Frishman D, Hayer-Hartl M, Mann M, Hartl FU (2005) Proteome-wide analysis of chaperonin-dependent protein folding in *Escherichia coli*. Cell 122(2):209–220

Khalil AA, Kabapy NF, Deraz SF, Smith C (2011) Heat shock proteins in oncology: diagnostic biomarkers or therapeutic targets? Biochim Biophys Acta 1816(2):89–104

Kim SW, Kim JB, Kim JH, Lee JK (2007) Interferon-gamma-induced expressions of heat shock protein 60 and heat shock protein 10 in C6 astroglioma cells: identification of the signal transducers and activators of transcription 3-binding site in bidirectional promoter. Neuroreport 18(4):385–389

Kirchhoff SR, Gupta S, Knowlton AA (2002) Cytosolic heat shock protein 60, apoptosis, and myocardial injury. Circulation 105(24):2899–2904

Kleinridders A, Lauritzen HP, Ussar S, Christensen JH, Mori MA, Bross P, Kahn CR (2013) Leptin regulation of Hsp60 impacts hypothalamic insulin signaling. J Clin Invest 123 (11):4667–4680

Kmiecik S, Kolinski A (2011) Simulation of chaperonin effect on protein folding: a shift from nucleation-condensation to framework mechanism. J Am Chem Soc 133(26):10283–10289

Koster KL, Sturm M, Herebian D, Smits SH, Spiekerkoetter U (2014) Functional studies of 18 heterologously expressed medium-chain acyl-CoA dehydrogenase (MCAD) variants. J Inherit Metab Dis 37(6):917–928

Koumoto Y, Shimada T, Kondo M, Takao T, Shimonishi Y, Hara-Nishimura I, Nishimura M (1999) Chloroplast Cpn20 forms a tetrameric structure in *Arabidopsis thaliana*. Plant J 17 (5):467–477

Kurochkina LP, Semenyuk PI, Orlov VN, Robben J, Sykilinda NN, Mesyanzhinov VV (2012) Expression and functional characterization of the first bacteriophage-encoded chaperonin. J Virol 86(18):10103–10111

Lalonde R, Strazielle C (2011) Brain regions and genes affecting limb-clasping responses. Brain Res Rev 67(1–2):252–259

Langer T (2000) AAA proteases: cellular machines for degrading membrane proteins. Trends Biochem Sci 25(5):247–251

Leitner A, Joachimiak LA, Bracher A, Monkemeyer L, Walzthoeni T, Chen B, Pechmann S, Holmes S, Cong Y, Ma B, Ludtke S, Chiu W, Hartl FU, Aebersold R, Frydman J (2012) The molecular architecture of the eukaryotic chaperonin TRiC/CCT. Structure 20(5):814–825

Levy-Rimler G, Bell R, Ben Tal N, Azem A (2002) Type I chaperonins: not all are created equal. FEBS Lett 529(1):1

Li Y, Malkaram SA, Zhou J, Zempleni J (2014) Lysine biotinylation and methionine oxidation in the heat shock protein HSP60 synergize in the elimination of reactive oxygen species in human cell cultures. J Nutr Biochem 25(4):475–482

Lianos GD, Alexiou GA, Mangano A, Mangano A, Rausei S, Boni L, Dionigi G, Roukos DH (2015) The role of heat shock proteins in cancer. Cancer Lett. doi:10.1016/j.canlet.2015.02.026

Lin L, Kim SC, Wang Y, Gupta S, Davis B, Simon SI, Torre-Amione G, Knowlton AA (2007) HSP60 in heart failure: abnormal distribution and role in cardiac myocyte apoptosis. Am J Physiol Heart Circ Physiol 293(4):H2238–H2247

Lopez T, Dalton K, Frydman J (2015) The mechanism and function of group II chaperonins. J Mol Biol. doi:10.1016/j.jmb.2015.04.013

Lorimer GH (1996) A quantitative assessment of the role of the chaperonin proteins in protein folding in vivo. FASEB J 10(1):5–9

Lund P (2011) Insights into chaperonin function from studies on archaeal thermosomes. Biochem SocTrans 39(1):94–98

Lund PA (2009) Multiple chaperonins in bacteria—why so many? FEMS Microbiol Rev 33 (4):785–800

Luo Y, Chen C, Zhan Z, Wang Y, Du J, Hu Z, Liao X, Zhao G, Wang J, Yan X, Jiang H, Pan Q, Xia K, Tang B, Shen L (2014) Mutation and clinical characteristics of autosomal-dominant hereditary spastic paraplegias in China. Neurodegener Dis 14(4):176–183

Magen D, Georgopoulos C, Bross P, Ang D, Segev Y, Goldsher D, Nemirovski A, Shahar E, Ravid S, Luder A, Heno B, Gershoni-Baruch R, Skorecki K, Mandel H (2008) Mitochondrial hsp60 chaperonopathy causes an autosomal-recessive neurodegenerative disorder linked to brain hypomyelination and leukodystrophy. Am J Hum Genet 83(1):30–42

Magnoni R, Palmfeldt J, Christensen JH, Sand M, Maltecca F, Corydon TJ, West M, Casari G, Bross P (2013) Late onset motoneuron disorder caused by mitochondrial Hsp60 chaperone deficiency in mice. Neurobiol Dis 54:12–23

Magnoni R, Palmfeldt J, Hansen J, Christensen JH, Corydon TJ, Bross P (2014) The Hsp60 folding machinery is crucial for manganese superoxide dismutase folding and function. Free Radic Res 48(2):168–179

Maier EM, Gersting SW, Kemter KF, Jank JM, Reindl M, Messing DD, Truger MS, Sommerhoff CP, Muntau AC (2009) Protein misfolding is the molecular mechanism underlying MCADD identified in newborn screening. Hum Mol Genet 18(9):1612–1623

Makino S, Whitehead GG, Lien CL, Kim S, Jhawar P, Kono A, Kawata Y, Keating MT (2005) Heat-shock protein 60 is required for blastema formation and maintenance during regeneration. Proc Natl Acad Sci USA 102(41):14599–14604

Martin CC, Tsang CH, Beiko RG, Krone PH (2002) Expression and genomic organization of the zebrafish chaperonin gene complex. Genome 45(5):804–811

Martinez A, Calvo AC, Teigen K, Pey AL (2008) Rescuing proteins of low kinetic stability by chaperones and natural ligands phenylketonuria, a case study. Prog Mol Biol Transl Sci 83: 89–134

Mclennan NF, Girshovich AS, Lissin NM, Charters Y, Masters M (1993) The strongly conserved carboxyl-terminus glycine methionine motif of the *Escherichia-Coli* groel chaperonin is dispensable. Mol Microbiol 7(1):49–58

McMullin TW, Hallberg RL (1987) A normal mitochondrial protein is selectively synthesized and accumulated during heat shock in *Tetrahymena thermophila*. Mol Cell Biol 7(12):4414–4423

McMullin TW, Hallberg RL (1988) A highly evolutionarily conserved mitochondrial protein is structurally related to the protein encoded by the *Escherichia coli* groEL gene. Mol Cell Biol 8 (1):371–380

Merendino AM, Bucchieri F, Campanella C, Marciano V, Ribbene A, David S, Zummo G, Burgio G, Corona DF, de Conway ME, Macario AJ, Cappello F (2010) Hsp60 is actively secreted by human tumor cells. PLoS One 5(2):e9247

Mitraki A, Danner M, King J, Seckler R (1993) Temperature-sensitive mutations and 2nd-site suppressor substitutions affect folding of the P22 tailspike protein in vitro. J Biol Chem 268:20071–20075

Mori D, Nakafusa Y, Miyazaki K, Tokunaga O (2005) Differential expression of Janus kinase 3 (JAK3), matrix metalloproteinase 13 (MMP13), heat shock protein 60 (HSP60), and mouse double minute 2 (MDM2) in human colorectal cancer progression using human cancer cDNA microarrays. Pathol Res Pract 201(12):777–789

Nargund AM, Fiorese CJ, Pellegrino MW, Deng P, Haynes CM (2015) Mitochondrial and nuclear accumulation of the transcription factor ATFS-1 promotes OXPHOS recovery during the UPR (mt). Mol Cell 58(1):123–133

Nargund AM, Pellegrino MW, Fiorese CJ, Baker BM, Haynes CM (2012) Mitochondrial import efficiency of ATFS-1 regulates mitochondrial UPR activation. Science 337(6094):587–590

Nisemblat S, Parnas A, Yaniv O, Azem A, Frolow F (2014) Crystallization and structure determination of a symmetrical 'football' complex of the mammalian mitochondrial Hsp60-Hsp10 chaperonins. Acta Crystallogr Sect F Struct Biol Commun 70(Pt 1):116–119

Nisemblat S, Yaniv O, Parnas A, Frolow F, Azem A (2015) Crystal structure of the human mitochondrial chaperonin symmetrical football complex. Proc Natl Acad Sci USA

Niwa T, Kanamori T, Ueda T, Taguchi H (2012) Global analysis of chaperone effects using a reconstituted cell-free translation system. Proc Natl Acad Sci USA 109(23):8937–8942

Niwa T, Ying BW, Saito K, Jin W, Takada S, Ueda T, Taguchi H (2009) Bimodal protein solubility distribution revealed by an aggregation analysis of the entire ensemble of *Escherichia coli* proteins. Proc Natl Acad Sci USA 106(11):4201–4206

Noreau A, Dion PA, Rouleau GA (2014) Molecular aspects of hereditary spastic paraplegia. Exp Cell Res 325(1):18–26

Nunnari J, Suomalainen A (2012) Mitochondria: in sickness and in health. Cell 148(6):1145–1159

O'Reilly L, Bross P, Corydon TJ, Olpin SE, Hansen J, Kenney JM, McCandless SE, Frazier DM, Winter V, Gregersen N, Engel PC, Andresen BS (2004) The Y42H mutation in medium-chain acyl-CoA dehydrogenase, which is prevalent in babies identified by MS/MS-based newborn screening, is temperature sensitive. Eur J Biochem 271(20):4053–4063

Parnas A, Nadler M, Nisemblat S, Horovitz A, Mandel H, Azem A (2009) The MitCHAP-60 disease is due to entropic destabilization of the human mitochondrial Hsp60 oligomer. J Biol Chem 284(41):28198–28203

Parnas A, Nisemblat S, Weiss C, Levy-Rimler G, Pri-Or A, Zor T, Lund PA, Bross P, Azem A (2012) Identification of elements that dictate the specificity of mitochondrial hsp60 for its co-chaperonin. PLoS One 7(12):e50318

Pelham HR (1986) Speculations on the functions of the major heat shock and glucose-regulated proteins. Cell 46(7):959–961

Peng L, Fukao Y, Myouga F, Motohashi R, Shinozaki K, Shikanai T (2011) A chaperonin subunit with unique structures is essential for folding of a specific substrate. PLoS Biol 9(4):e1001040

Perezgasga L, Segovia L, Zurita M (1999) Molecular characterization of the 5' control region and of two lethal alleles affecting the hsp60 gene in *Drosophila melanogaster*. FEBS Lett 456 (2):269–273

Pey AL, Desviat LR, Gamez A, Ugarte M, Perez B (2003) Phenylketonuria: genotype-phenotype correlations based on expression analysis of structural and functional mutations in PAH. Hum Mutat 21(4):370–378

Pockley AG, Henderson B, Multhoff G (2014) Extracellular cell stress proteins as biomarkers of human disease. Biochem Soc Trans 42(6):1744–1751

Powers ET, Balch WE (2013) Diversity in the origins of proteostasis networks—a driver for protein function in evolution. Nat Rev Mol Cell Biol 14(4):237–248

Racis L, Di Fabio R, Tessa A, Guillot F, Storti E, Piccolo F, Nesti C, Tedde A, Pierelli F, Agnetti V, Santorelli FM, Casali C (2014) Large deletion mutation of SPAST in a multi-generation family from Sardinia. Eur J Neurol 21(6):935–938

Rao VB, Black LW (2010) Structure and assembly of bacteriophage T4 head. Virol J 7:356. doi:10.1186/1743-422x-7-356

Richardson A, Landry SJ, Georgopoulos C (1998) The ins and outs of a molecular chaperone machine. Trends Biochem Sci 23(4):138–143

Richardson A, Schwager F, Landry SJ, Georgopoulos C (2001) The importance of a mobile loop in regulating chaperonin/ co-chaperonin interaction: humans versus *Escherichia coli*. J Biol Chem 276(7):4981–4987

Richardson A, van der Vies SM, Keppel F, Taher A, Landry SJ, Georgopoulos C (1999) Compensatory changes in GroEL/Gp31 affinity as a mechanism for allele- specific genetic interaction. J Biol Chem 274(1):52–58

Rinaldo P, Matern D, Bennett MJ (2002) Fatty acid oxidation disorders. Annu Rev Physiol 64:477–502

Rospert S, Junne T, Glick BS, Schatz G (1993) Cloning and disruption of the gene encoding yeast mitochondrial chaperonin 10, the homolog of *E. coli* groES. FEBS Lett 335(3):358–360

Rutherford SL (2000) From genotype to phenotype: buffering mechanisms and the storage of genetic information. BioEssays 22(12):1095–1105

Rutherford SL (2003) Between genotype and phenotype: protein chaperones and evolvability. Nat Rev Genet 4(4):263–274

Rutherford SL, Lindquist S (1998) Hsp90 as a capacitor for morphological evolution. Nature 396 (6709):336–342

Ryan MT, Herd SM, Sberna G, Samuel MM, Hoogenraad NJ, Hoj PB (1997) The genes encoding mammalian chaperonin 60 and chaperonin 10 are linked head-to-head and share a bidirectional promoter. Gene 196(1–2):9–17

Saccoccia F, Di MP, Boumis G, Brunori M, Koutris I, Miele AE, Morea V, Sriratana P, Williams DL, Bellelli A, Angelucci F (2012) Moonlighting by different stressors: crystal structure of the chaperone species of a 2-Cys peroxiredoxin. Structure 20(3):429–439

Saibil H (2013) Chaperone machines for protein folding, unfolding and disaggregation. Nat Rev Mol Cell Biol 14(10): 630–642

Saibil HR, Fenton WA, Clare DK, Horwich AL (2013) Structure and allostery of the chaperonin GroEL. J Mol Biol 425(9):1476–1487

Saijo T, Welch WJ, Tanaka K (1994) Intramitochondrial folding and assembly of medium-chain acyl-CoA dehydrogenase (MCAD)—demonstration of impaired transfer of K304E-variant MCAD from its complex with Hsp60 to the native tetramer. J Biol Chem 269:4401–4408

Sakamoto M, Ohkuma M (2010) Usefulness of the hsp60 gene for the identification and classification of Gram-negative anaerobic rods. J Med Microbiol 59(Pt 11):1293–1302

Sakikawa C, Taguchi H, Makino Y, Yoshida M (1999) On the maximum size of proteins to stay and fold in the cavity of GroEL underneath GroES. J Biol Chem 274(30):21251–21256

Schumann W (1996) Regulation of the heat shock response in *Escherichia coli* and Bacillus subtilis. J Biosci 21(2):133–148

Siegers K, Waldmann T, Leroux MR, Grein K, Shevchenko A, Schiebel E, Hartl FU (1999) Compartmentation of protein folding in vivo: sequestration of non- native polypeptide by the chaperonin-GimC system. EMBO J 18(1):75–84

Sorensen HP, Mortensen KK (2005) Advanced genetic strategies for recombinant protein expression in *Escherichia coli*. J Biotechnol 115(2):113–128

Stan G, Lorimer GH, Thirumalai D, Brooks BR (2007) Coupling between allosteric transitions in GroEL and assisted folding of a substrate protein. Proc Nat Acad Sci USA 104(21):8803–8808

Stemp MJ, Guha S, Hartl FU, Barral JM (2005) Efficient production of native actin upon translation in a bacterial lysate supplemented with the eukaryotic chaperonin TRiC. Biol Chem 386(8):753–757

Studer RA, Dessailly BH, Orengo CA (2013) Residue mutations and their impact on protein structure and function: detecting beneficial and pathogenic changes. Biochem J 449(3): 581–594

Suzuki K, Nakanishi H, Bower J, Yoder DW, Osteryoung KW, Miyagishima SY (2009) Plastid chaperonin proteins Cpn60 alpha and Cpn60 beta are required for plastid division in *Arabidopsis thaliana*. BMC Plant Biol 9:38

Svenstrup K, Bross P, Koefoed P, Hjermind LE, Eiberg H, Born AP, Vissing J, Gyllenborg J, Norremolle A, Hasholt L, Nielsen JE (2009) Sequence variants in SPAST, SPG3A and HSPD1 in hereditary spastic paraplegia. J Neurol Sci 284(1–2):90–95

Takano T, Kakefuda T (1972) Involvement of a bacterial factor in morphogenesis of bacteriophage capsid. Nat New Biol 239(89):34–37

Tang YC, Chang HC, Roeben A, Wischnewski D, Wischnewski N, Kerner MJ, Hartl FU, Hayer-Hartl M (2006) Structural features of the GroEL-GroES nano-cage required for rapid folding of encapsulated protein. Cell 125(5):903–914

Techtmann SM, Robb FT (2010) Archaeal-like chaperonins in bacteria. Proc Natl Acad Sci USA 107(47):20269–20274

Tian GL, Vainberg IE, Tap WD, Lewis SA, Cowan NJ (1995) Specificity in chaperonin-mediated protein folding. Nature 375:250–253

Todd MJ, Lorimer GH, Thirumalai D (1996) Chaperonin-facilitated protein folding: optimization of rate and yield by an iterative annealing mechanism. Proc Natl Acad Sci USA 93(9): 4030–4035

Tokuriki N, Tawfik DS (2009) Chaperonin overexpression promotes genetic variation and enzyme evolution. Nature 459(7247):668–673

Trent JD, Nimmesgern E, Wall JS, Hartl FU, Horwich AL (1991) A molecular chaperone from a thermophilic archaebacterium is related to the eukaryotic protein t-complex polypeptide-1. Nature 354(6353):490–493

van der Vies SM, Gatenby AA, Georgopoulos C (1994) Bacteriophage-T4 encodes a co-chaperonin that can substitute for *Escherichia coli* GroES in protein folding. Nature 368:654–656

Van Dyk TK, Gatenby AA, LaRossa RA (1989) Demonstration by genetic suppression of interaction of GroE products with many proteins. Nature 342(6248):451–453

Vendruscolo M, Knowles TP, Dobson CM (2011) Protein solubility and protein homeostasis: a generic view of protein misfolding disorders. Cold Spring Harb Perspect Biol 3(12):a010454

Vihervaara A, Sistonen L (2014) HSF1 at a glance. J Cell Sci 127(Pt 2):261–266

Viitanen PV, Gatenby AA, Lorimer GH (1992) Purified chaperonin 60 (groEL) interacts with the nonnative states of a multitude of *Escherichia coli* proteins. Protein Sci 1(3):363–369

Villebeck L, Moparthi SB, Lindgren M, Hammarstrom P, Jonsson BH (2007) Domain-specific chaperone-induced, expansion is required for beta-actin folding: a comparison of beta-actin conformations upon interactions with GroEL and tail-less complex polypeptide 1 ring complex (TRiC). Biochemistry-US 46(44):12639–12647

Viscomi C, Bottani E, Zeviani M (2015) Emerging concepts in the therapy of mitochondrial disease. Biochim Biophys Acta 6–7:544–557

Vitlin Gruber A, Nisemblat S, Azem A, Weiss C (2013) The complexity of chloroplast chaperonins. Trends Plant Sci 18(12):688–694

Vitlin Gruber A, Zizelski G, Azem A, Weiss C (2014) The Cpn10(1) co-chaperonin of *A. thaliana* functions only as a hetero-oligomer with Cpn20. PLoS One 9(11):e113835

Wakano C, Byun JS, Di LJ, Gardner K (2012) The dual lives of bidirectional promoters. Biochim Biophys Acta 7:688–693

Warnecke T, Hurst LD (2010) GroEL dependency affects codon usage–support for a critical role of misfolding in gene evolution. Mol Syst Biol 6:340

Williams TA, Fares MA (2010) The effect of chaperonin buffering on protein evolution. Genome Biol Evol 2:609–619

Wyganowski KT, Kaltenbach M, Tokuriki N (2013) GroEL/ES buffering and compensatory mutations promote protein evolution by stabilizing folding intermediates. J Mol Biol 425 (18):3403–3414

Xu Z, Sigler PB (1998) GroEL/GroES: structure and function of a two-stroke folding machine. J Struct Biol 124(2–3):129–141

Xu ZH, Horwich AL, Sigler PB (1997) The crystal structure of the asymmetric GroEL-GroES-(ADP)(7) chaperonin complex. Nature 388(6644):741–750

Yam AY, Xia Y, Lin HT, Burlingame A, Gerstein M, Frydman J (2008) Defining the TRiC/CCT interactome links chaperonin function to stabilization of newly made proteins with complex topologies. Nat Struct Mol Biol 15(12):1255–1262

Yang D, Ye X, Lorimer GH (2013) Symmetric GroEL: GroES2 complexes are the protein-folding functional form of the chaperonin nanomachine. Proc Natl Acad Sci USA 110(46):E4298–E4305

Ye X, Lorimer GH (2013) Substrate protein switches GroE chaperonins from asymmetric to symmetric cycling by catalyzing nucleotide exchange. Proc Natl Acad Sci USA 110(46): E4289–E4297

Yebenes H, Mesa P, Munoz IG, Montoya G, Valpuesta JM (2011) Chaperonins: two rings for folding. Trends Biochem Sci 36(8):424–432

Yogev O, Pines O (2011) Dual targeting of mitochondrial proteins: mechanism, regulation and function. Biochim Biophys Acta 3:1012–1020

Yokota I, Saijo T, Vockley J, Tanaka K (1992) Impaired tetramer assembly of variant medium-chain acyl-coenzyme a dehydrogenase with a glutamate or aspartate substitution for lysine 304 causing instability of the protein. J Biol Chem 267:26004–26010

Yu H, Lee H, Herrmann A, Buettner R, Jove R (2014) Revisiting STAT3 signalling in cancer: new and unexpected biological functions. Nat Rev Cancer 14(11):736–746

Zahn R, Lindner P, Axmann SE, Pluckthun A (1996) Effect of single point mutations in citrate synthase on binding to GroEL. FEBS Lett 380(1–2):152–156

Zanin-Zhorov A, Cahalon L, Tal G, Margalit R, Lider O, Cohen IR (2006) Heat shock protein 60 enhances CD4CD25 regulatory T cell function via innate TLR2 signaling. J Clin Invest 116 (7):2022

Zeilstra-Ryalls J, Fayet O, Georgopoulos C (1991) The universally conserved GroE (Hsp60) chaperonins. Annu Rev Microbiol 45:301–325

Zhang Q, Nishimura D, Vogel T, Shao J, Swiderski R, Yin T, Searby C, Carter CS, Kim G, Bugge K, Stone EM, Sheffield VC (2013) BBS7 is required for BBSome formation and its absence in mice results in Bardet-Biedl syndrome phenotypes and selective abnormalities in membrane protein trafficking. J Cell Sci 126(Pt 11):2372–2380

Zhang Y, Iqbal S, O'Leary MF, Menzies KJ, Saleem A, Ding S, Hood DA (2013) Altered mitochondrial morphology and defective protein import reveal novel roles for Bax and/or Bak in skeletal muscle. Am J Physiol Cell Physiol 305(5):C502–C511

Zhao Q, Wang J, Levichkin IV, Stasinopoulos S, Ryan MT, Hoogenraad NJ (2002) A mitochondrial specific stress response in mammalian cells. EMBO J 21(17):4411–4419

Printed in the United States
By Bookmasters